Textbook on Food Science and Human Nutrition
for Undergraduates

THE AUTHOR

Dipti Sharma is an Assistant Professor in the Department of Food Technology at a College of University of Delhi. She has her specialization in Food Science and Technology & have completed B.A. Sc. & M.Sc. in Food Technology from University of Delhi and G.J.U of Science and Technology respectively. She is pursuing Ph.D in Food Technology from Dr. K.N. Modi University, Rajasthan. She has completed a Certificate Course in Consumer Protection and a PG Diploma in Food Safety and Quality Management from IGNOU. She has more than 8 years of industry and teaching experience.

She is a life member of the Association of Food Scientist and Technologists of India, Society of Indian Bakers and ISTE. Besides coordinating a number of workshops and seminars, she has presented papers at the national and international conferences. She is also in the reviewer/editorial board of journal.

Textbook on Food Science and Human Nutrition
for Undergraduates

Dipti Sharma

M.Sc. and B.A.Sc. (Food Technology)
Assistant Professor
University of Delhi

2015

Daya Publishing House®
A Division of
Astral International Pvt. Ltd.
New Delhi – 110 002

ISBN 9789351302889 (International Edition)

Published by	: **Daya Publishing House®** A Division of **Astral International Pvt. Ltd.** – ISO 9001:2008 Certified Company – 4760-61/23, Ansari Road, Darya Ganj New Delhi-110 002 Ph. 011-43549197, 23278134 E-mail: info@astralint.com Website: www.astralint.com
Laser Typesetting	: **Classic Computer Services**, Delhi - 110 035
Printed at	: **Thomson Press India Limited**

PRINTED IN INDIA

I dedicated this book to

my Guruji "Shri Ashutosh Maharajji",

my Son
to my Mother for continuous inspiration
to my Father & Brothers for their support and
to my loving Husband
for constant encouragement and motivation

Acknowledgement

I express my gratitude to the scientists and researchers whose textbooks, journals, articles of national and international origin have been studied for writing this textbook. I am also thankful to the National Institute of Nutrition, Hyderabad and ICMR whose research findings and RDA's have been incorporated in the book for information dissemination purpose to students and updating teachers with the latest findings.

I am thankful to Astral International (Pvt) Ltd. for accepting and publishing the book.

Dipti Sharma

Foreword

I am very happy to learn that a book titled *"Textbook on Food & Nutrition for undergraduates"*, written by Dipti Sharma, Assistant Professor, University of Delhi is being published. This book covers the subject of Basic & Applied Nutrition for students of Food Technology, Food Science & Technology, Food & Nutrition courses at undergraduate level taught in Indian Universities. The topics are dealt in detail with particular emphasis to Indian context.

This well written, concise information apart from being an excellent academic tool, it shall also find extensive application in the food industry as a guide for training the basics of nutrition.

I, on behalf of Subharti University, congratulate everyone associated with the publication of this book and extend best wishes for successful publication of the book.

Prof. (Dr.) Ritu Baksi
Head, Department of Home Science
Subharti University
Meerut

Preface

This text book is a result of my teaching at one of the college of University of Delhi, absence of single text book which has all the required information has motivated myself to write the " Textbook on Food & Nutrition for undergraduates". This book covers the subject of Basic & Applied Nutrition in Food Technology, Food Science & Technology, Food & Nutrition courses at undergraduate level taught in Indian Universities. I am appreciative of student queries and of those who provided materials used throughout book to offer better explanations of the text. The topics are dealt in detail with particular emphasis to Indian context.

This book gives revised Recommended Dietary Allowances suggested by Indian Council of Medical Research in year 2010. In this book special emphasis is given to food constituents, Nutrients and Labelling. The book contains the FSSAI guidelines on labelling. This book can serve as textbook for undergraduates studying Food and Nutrition course. This book can also be used by postgraduates and students doing paramedical courses like nursing. This book would also serve as a valuable reference source for teachers.

The book is divided into three parts: Unit I- Introduction to Food and Nutrition, Unit II- Nutrients, Unit III – deals with Effect of cooking on food, Nutritional improvement of food and Labelling.

I hope that readers find the book more reader-friendly and more useful in explaining food science and nutrition concepts. Feedback and suggestions from colleagues as well as students to make the book more useful are most welcome.

Dipti Sharma
Assistant Professor
University of Delhi

Contents

Unit I

1

Introduction to Food and Nutrition

1.0 Introduction and History of Nutrition

Our mother nature has provided a wide variety of foods for man to consume and stay healthy. We consume food for the maintenance of good health, growth and development and to protect against infections. Foods contain number of substances called nutrients and non nutritive factors in varying proportions. Understanding the functions of these nutrients and major food sources is important for man to formulate a nutritious diet.

The science of nutrition has been developed by using the combined knowledge of the physical and biological sciences. Its application involves the social sciences related to man's behaviour - Psychology, sociology, anthropology and economics. Until World War I the significance of nutrition was recognized by a relatively small group of scientists and physicians. Since then, a wider understanding has been developed on the role of food and nutrients in maintaining health of individuals and the economic development of the nation.

Some research in the field of nutrition were stimulated by national emergencies. Others depended on technical development of the supporting sciences. Nutrition research in India, as beri – beri inquiry was started in 1918, under the guidance of Sir Mc Carrison at Coonoor in South India. It is now an important national institution, at Hyderabad and is known as National Institute of Nutrition. The institute is currently engaged in carrying out basic as well as applied research work in nutrition. This national institute comes under the Indian Council of Medical Research (ICMR).

1.1 Definition

☆ **Food Science:** Food is a mixture of many different chemical components. The study of food science involves understanding the reactions which may be a result of the interaction between components, with the cooking medium and the processing/environmental conditions like heat, cold, light and air to which the food is subjected during food preparation, preservation etc. Many physical and chemical reactions occur during food preparation. The study of food science also includes understanding the nutritive value of different foods and methods of preserving them during cooking.

☆ **Food Technology** is the application of principles of food science and engineering to the process and preserves food and give convenience to consumer.

☆ **Nutrition** is defined as a science concerned with the role of food and nutrients in the maintenance of health. Nutrition as defined by Robinson (1982) is "the science of foods and nutrients, their action, interaction and balance in relationship to health and disease, the processes by which the organism ingests, digests, absorbs, transports and utilizes nutrients and disposes of their end product".

The council on food and nutrition of the American Medical Association defines Nutrition as "the science of food, the nutrient and the substances therein, their action, interaction and balance in relation to health and disease and the process by which the organism ingests, digests, absorbs, transports, utilizes and excretes waste substances".

☆ **Food** refers to whatever is eatable and is the source of nutrients for the body.

☆ **Nutrients** are the nourishing substances in food that are essential for growth of the infant, development from childhood to adulthood, and the maintenance of bodily function throughout life. These are the constituents in food that must be supplied to the body in adequate amounts. These include carbohydrates, proteins, fats, minerals and vitamins.

In nutrition, an essential nutrient is one whose omission from the diet will lead to a decline in certain aspects of human health such as function of the heart etc. If the omitted nutrient is not restored in the body either by diet or supplement then it may lead to permanent damage of human health or impairment of body's normal function. Essential nutrients of human diet and their classes are shown below:

Energy Yielding Nutrients			Water
Carbohydrate	Fats	Proteins	
Glucose	Linoleic acid (ω-6), Linolenic acid (ω-3)	Isoleucine, Histidine, Leucine, Lysine, Threonine, Valine, Tryptophan, Methionine, Phenylalanine	Water

Protective and Regulatory Nutrients			
Vitamins		*Minerals*	
Fat Soluble Vitamins	*Water Soluble Vitamins*	*Major*	*Trace*
Vitamin A, D, E, and K	Thiamine, Niacin, Riboflavin, Biotin, Vitamin B6, B12, Pantothenic acid	Calcium, Chloride, Magnesium, Phosphorus, Potassium, Sodium, Sulphur	Copper, Zinc, Chromium, Iron, Fluoride, Selenium Manganese, Iodide, Molybdenum

Although dietary fiber are essential for maintaining good health but are not considered in the list of essential substance but it is not a nutrient. Dietary fiber passes through the small intestine undigested to provide bulk for stool.

☆ **Nutritional status** is the condition of health of the individual as influenced by the utilization of the nutrients.

☆ **Nutrient Requirement** is defined as the minimum amount of the absorbed nutrient that is necessary for maintaining the normal physiological functions of the body.

☆ **Bioavailability** : The term bioavailability indicates what is absorbed and utilized by the body.

☆ **Dietary Reference Intakes (DRI):** In America the term RDA is being replaced by DRI which refers to Dietary Reference Intakes. It provide information on following four aspects:

1. **Recommended Dietary Allowances (RDA)** are the requirements of the quantities of any nutrients that healthy individuals must obtain from food to meet their physiological needs. The recommended dietary allowances (RDAs) are estimates of nutrients to be consumed daily to ensure the requirements of all individuals in a given population. It is defined as estimated nutrient allowance that is adequate in 97-98 per cent of the healthy population specific for life-stage, age, gender and physical activity level. The recommended level depends upon the bioavailability of nutrients from a given diet. RDA includes addition of safety factor or margin to the requirement of the nutrient, to cover the variation among individuals, dietary traditions, practices, losses during cooking and the lack of precision inherent in the estimated requirement. The RDA is the dietary intake goal for individuals, but its purpose is not to assess diets of individuals or groups but to provide information to the individual how much nutrient he must eat to maintain good health. The RDA's are suggested for physiological groups such as infants, pre-schoolers, children, adolescents, pregnant women, lactating mothers, and adult men and women, taking into account their physical activity. In fact, RDAs are suggested averages/day. However, in practice, fluctuations in intake may occur depending on the food availability and demands of the body. But, the average requirements need to be satisfied over a period of time

2. **Estimated Average Requirements (EAR)** is the estimated nutrient requirement that is adequate in 50 per cent of the population. This may be used to assess diets of individuals or groups and is used to develop the RDA's.

3. **Tolerable Upper Intake Level (UL)** refers to the maximum nutrient intake that is not associated with adverse side effects in most individuals of a healthy population.

4. **Adequate Intake (AI)** is used when insufficient scientific evidence exits to calculate the EAR and RDA, and may be used as a goal for dietary intakes of individual.

☆ **Antioxidants:** antioxidants as "compounds that protect biological systems against the potentially harmful effects of processes or reactions that can cause extensive oxidations." The ultimate purpose of food antioxidants is to inhibit oxidative reactions that cause deterioration of quality (*e.g.,* of flavor, colour, nutrient composition, texture). With this goal in mind, food antioxidants can be defined as any compounds serving to inhibit oxidative processes that can deteriorate the quality of food lipids. Antioxidant prevent the deterioration of food lipid by free radical scavenging, inactivation of peroxides and other reactive oxygen species, chelation of metals, and quenching of secondary lipid oxidation products that produce rancid odours. Antioxidants also help control the activity of transition metals, singlet oxygen and enzymes.

☆ **Phytochemicals:** are non-nutrient compounds found in plant derived food that has biological activity in the body. These chemicals of plant origin include terpenes, phytosterols, flavanoids, theols and allylic sulphides which are antimutagenic and anticarcinogenic agents and thus render nutraceutical properties. More than 5000 phytochemicals have been identified in plant foods and many more remain to be discovered. Carotenoids are phytochemicals that are abundantly present in yellow-orange coloured fruits and vegetables and their increased consumption is associated with several health benefits. Nuts contain many phytochemicals (flavonoids, phenolic compounds, isoflavones, sterols, etc.), which have been shown to be inversely related to CHD.

☆ **Health-promoting foods** or compounds are generally classified into two major categories:

1. **Neutraceutical**: In the year 1989 the word "nutraceutical," is a blend of "nutrition" and "pharmaceutical" and was coined by Dr. Stephen De Felice, a physician who founded the Foundation for Innovation in Medicine, USA. At that time, Dr. De Felice defined "nutraceutical" as "any food or parts of a food that provides medical or health benefits, including the prevention and treatment of diseases".

Nutraceuticals are the health promoting compounds or products that have been isolated or purified from food sources and they are generally sold in a medicinal (usually pill) form. A good example is a group of compounds called isoflavones that are isolated from soybean seeds and packaged into pills that women can use instead of synthetic compounds during hormone replacement therapy. Other examples of nutraceutical products include fish oil like cod liver oil capsules, herb extracts, glucosamine, lutein-containing multivitamin tablets, and antihypertensive pills that contain fish protein-derived peptides.

2. **Functional Food:** Functional foods are the products that may look like or be a conventional food and be consumed as part of a usual diet, but apart from supplying nutrients they can reduce the risk of chronic diseases such as cancer, hypertension, kidney malfunction etc. A typical example of a functional food is tomato fruit which contain specific type of compound (carotenoid) that helps to remove toxic compounds from our body and thereby prevent damage to essential organs like the heart, kidney, lungs, brain, etc. Other typical examples of functional foods include soybean, fish, oat meal, cereal bran (wheat, rice), and tea (green and black). Apart from traditional foods, there are also functional foods that are produced through food processing such as the antihypertensive sour milk that has been shown to reduce blood pressure in human beings. Functional foods are added to certain foods or food products with purpose of disease prevention and improved health benefits.

Development and regulatory interest in functional foods began in Japan in the early eighties with advances in chemical identification of bioactive compounds, processing and formulation of foods as well as elucidation of molecular mechanisms involved in the modulation of metabolic disorders. The initial regulatory environment for functional foods was established by Japan in 1991 with the introduction of "Foods for Specified Health Use" (FOSHU) policy that enabled production and marketing of health-promoting foods. Since 1991 over 600 FOSHU products are now available in the Japanese market.

☆ **Basal Metabolic Rate (BMR):** BMR represents the minimum amount of energy expended in a fasting state (12 hours or more) to keep a resting, awake body alive in a warm, quiet environment. For a sedentary person, basal metabolism accounts for about 60 to 70 per cent of total energy expenditure. Some of the processes involved include the beating of the heart, respiration by the lungs, and the activity of other organs, such as the liver, brain, and kidney. It does not include energy expended for physical activity or the digestion, absorption, and processing of nutrients recently consumed.

☆ **Body Mass Index (BMI):** BMI is an excellent indicator of the weight status of any person. It is defined as the weight (in kg) divided by height (in m^2). As per an article on "Appropriate body-mass index for Asian populations and its implications for policy and intervention strategies published by WHO on Public health" the WHO expert consultation has concluded on the basis of the available data in Asia, that Asians generally have a higher percentage of body fat than white people of the same age, sex, and BMI.

Also, the proportion of Asian people with risk factors for type 2 diabetes and cardiovascular disease is substantial even below the existing WHO BMI cut-off point of 25 kg/m². Thus, current WHO cut-off points do not provide an adequate basis for taking action on risks related to overweight and obesity in many populations in Asia.

The purpose of a BMI cut-off point is to identify, within each population, the proportion of people with a high risk of an undesirable health state that warrants a public health or clinical intervention. When applied to a population, the purpose of anthropometric cut-off points is to identify independent and interactive risks of adverse health outcomes associated with different body compositions, so as to inform policy, trigger action, facilitate prevention programmes, and assess the effect of Interventions. Table 1.1 The following current WHO BMI cut-off points should be retained as international classification.

Table 1.1: BMI Cut-off Points for Asians

BMI	Classification
<16 kg/m²	Severe underweight
16·0–16·9 kg/m²	Moderate underweight
17·0–18·49 kg/m²	Mild underweight
18·5–24·9 kg/m²	Normal range
25 kg/m²	Overweight
25–29·9 kg/m²	Pre obese
30 kg/m²	Obesity
30–39·9 kg/m²	Obese class I
35–39·9 kg/m²	Obese class II
40 kg/m²	Obese class III

But the cut-off points of 23, 27·5, 32·5, and 37·5 kg/m² are to be added as points for public health action.

☆ **Balanced diet** is the diet which contains different types of food in such quantities and proportions so that the need for calories, proteins, minerals, vitamins and other nutrients is adequately met and a small provision is also made for extra nutrients to withstand short duration of leanness. The balanced diet should be prepared by keeping in mind the nutritional requirement of the individual so as to maintain good health. A balanced diet should be made by including food from all the food groups in a balanced amount so that the individual gets all types of food and nutrients to meet his or her requirement. Additionally balanced diet should provide bioactive phytochemicals such as dietary fibre, antioxidants and other nutraceuticals which have positive health benefits.

A balanced diet should provide around 60-70 per cent of total calories from carbohydrates, 20-25 per cent of total calories from fat and 10-12 per cent from protein. As the energy requirement is met from the major nutrient that doesn't mean that one should ignore including fruits and vegetables, emphasis should also be laid on including fruits and vegetables as they are source of vitamins and minerals that provides immunity to the body. Criteria for a balance diet:

☆ Should meet the nutritional requirements of the individual

☆ Should prevent degenerative diseases

☆ Should improves immunity

☆ Should increase stamina

☆ Should improves longevity

☆ Should helps in coping up stress

☆ Should develops optimum cognitive ability

☆ **Food additive** is defined as nutritive and non-nutritive substances added intentionally to food, generally in small quantities as per the quantity approved by FSSAI to improve its appearance, flavour, texture or storage properties.

☆ **Dietary or food supplements** refer to the concentrated sources of nutrient or other substances with a nutritional or physiological effect whose purpose is to supplement the normal diet.

1.2 Relation between Good Nutrition and Health

Health is defined by the World Health Organization (WHO) as the "State of complete physical, mental and social well being and not merely the absence of disease or infirmity". To maintain good health and nutritional status an individual must eat a balanced food containing all the nutrients in the adequate proportion from all the food groups. The prerequisites of health would include the following:

1. To achieve optimal growth and development.
2. To maintain structural integrity and functional efficiency of body tissues necessary for an active and productive use.
3. Mental well-being and emotional rest.
4. Ability to fight against diseases *i.e.* resisting infections by developing immune competence, preventing the onset of degenerative diseases like cancer and resisting the effect of environmental toxins/pollutants/allergens.
5. Ability to withstand the inevitable process of aging with minimal disability and functional impairment.

Earlier nutrition was thought be an important factor for proper growth, development and tissue integrity, but now with the development of society and changing lifestyle and eating habits of people the nutrition plays persuasive role in the other dimensions *i.e.* Health also. Hence an optimal nutritional status is an indication of good health. This recent advance has brought about a large-scale change in dietary habits and practices of the population. Table 1.2 gives nutrient deficiency symptoms and sources of nutrients and explains how food is related to health.

Table 1.2: Food in Relation to Maintain Health

Nutrient	Deficiency	Sources
Energy, Protein	Underweight, Kwashiorkor, Marasmus	Cereals, pulses, egg, meat fats and oils, sugar
Calcium	Tetany, Rickets, Osteomalacia	Milk, Cheese, Khoa, Gingelly seeds, Spinach, Okra, Soyabeans Some fish, like sardines, salmon, perch, and rainbow trout
Iron	Anaemia	Liver, green leaf vegetables, rice flakes, jaggery.
Vitamin-A	Night blindness	Liver, egg yolk, butter, green leafy vegetable, carrots
Vitamin-B12	Pernicious anaemia	Yeast, fermented foods.
Vitamin-D	Rickets, Osteomalacia	Sunlight, flesh foods–Beef liver, Fatty fish, like tuna, mackerel, and salmon, vitamin D dairy products, orange juice, soy milk, and cereals cheese, egg yolks
Vitamin-C	Scurvy, Bleeding gums	Citrus fruits, amla, guava, chilli
Folic acid	Megaloblastic anaemia	Fresh green leafy vegetables, lady's finger, cluster beans.
Vitamin-B6	Anaemia, angular stomatitis	Meat, liver, vegetables,whole cereal grains.
Niacin	Dementia, diarrhoea, dermatitis	Groundnuts, whole cereals, pulses
Thiamine	Pain in the calf muscle, weakness of heart muscle, defect in carbohydrate metabolism	Yeast, outer layers of cereals, pulses

Food can support good health is important. One example to support this given in the Table 1.3 which shows the clinical signs of a well nourished child as against those of an ill nourished child

Table 1.3

Signs of Well Nourished Child	Signs of Ill Nourished Child
1. Skin is smooth, pliable and elastic and of a healthy colour.	Lack of colour of skin – paleness
2. Bright and clear eyes and pink eye membranes.	Pale, dark red, or purple mucous membrane lining the eyes. Failing eye sight.
3. Firm pink nails.	Rigid brittle nails.
4. The hair is lustrous and firmly attached to the scalp.	Dull hair lacking sheen, dry, and can be easily plucked.
5. Healthy gums and membranes of the mouth.	Pale, dark red or purple colour of gums.
6. Reddish pink tongue. Not coated, pink lips.	Sores on skin, lip or tongue, pale lips.
7. Desirable height for age and desirable weight for height.	Stunted growth and weight deficit.

Contd...

Table 1.3–*Contd...*

Signs of Well Nourished Child	Signs of Ill Nourished Child
8. Good appetite and sound nutrition.	Loss of appetite, digestive disturbances, undernutrition.
9. Normal body temperature, pulse rate and breathing rate.	Above normal body temperature, shortness ofbreath while performing normal activity.
10. Healthy children are alert.	Listless, irritable and depressed.

Good Health of an individual can also be assessed by clinical assessment like blood test etc. The following Table 1.4 gives the clinical signs which arise due to deficiency of particular nutrient; this will help a physician and nutritionist in identifying the clinical signs. However, a physician must be consulted before taking any nutrient supplement to cure the deficiency.

Table 1.4: The Clinical Signs which Arise Due to Nutrient Deficiency

Condition	Clinical Signs
1. Protein Energy Malnutrition	Oedema, depigmentation, sparseness and hair fall, moon face, enlarged liver, muscle wasting.
2. Vitamin A deficiency (VAD)	Night blindness, Bitot's spots in the eye, Xerosis of skin.
3. Riboflavin deficiency	Angular stomatitis, cheilosis.
4. Thiamine deficiency	Oedema, sensory loss, calf muscle tenderness.
5. Niacin deficiency	Raw tongue, pigmentation of the skin.
6. Vitamin C deficiency	Spongy and bleeding gum.
7. Vitamin D deficiency	Rickets, beading of ribs, Knock – knees, bowed legs.
8. Iron deficiency	Pale conjunctiva, spoon shaped nails.
9. Iodine deficiency	Enlargement of thyroid gland.

How to stay Healthy ?

1. Exercise regularly.
2. Avoid smoking, chewing of tobacco and tobacco products (Khaini, Zarda, Paan masala) and consumption of alcohol.
3. Check regularly for blood sugar, lipids and blood pressure after the age of 30 years at least every 6 months.
4. Avoid self medication.
5. Adopt stress management techniques (Yoga and Meditation).

4.3 Concept of Malnutrition

Desirable Nutrition

refers to the nutritional status for a particular nutrient which is desirable when body tissues have enough of the nutrient to support normal metabolic function as well as surplus stores that can be metabolized in terms of increased needs.

Malnutrition as defined by World Health Organisation (WHO) is a pathological state resulting from a relative or absolute deficiency or excess of one or more essential nutrients, this state being clinically manifested or detected only by biochemical, anthropometric or physiological tests. Malnutrition includes both undernutrition and overnutrition.

   a. Undernutrition – the pathological state resulting from the consumption of an inadequate quantity of food over an extended period of time. When nutrient intake doesn't meet nutrient requirements, stores of nutrients soon become depleted by ongoing body use, some sooner than others, which results in undernutrition. The demand for these nutrients exists partly because the body is in a constant state of turnover. For example Marasmus is caused by undernutrition. Starvation implies total elimination of food and hence the rapid development of under nutrition and marasmus. Undernutrition leads to:

     1. Reduced biochemical functions: This will occur when tissue concentration of an essential nutrient falls below to support metabolic processes for example decreased enzyme function and reduced level of haemoglobin for iron synthesis in the body. This type of nutrient deficiency is termed as subclinical because there are not much outward sign or symptoms.

     2. Specific deficiency condition is the pathological state resulting from a relative or absolute lack of an individual nutrient. This type of nutrient deficiency is termed as clinical because there is the appearance of clinical sign or symptoms for example changes in skin, hair, nail, tongue colour etc.

   b. Over nutrition – Prolonged consumption of more nutrient than the body's requirement can lead to over nutrition. Example of over nutrition is obesity. Over nutrition may or may not cause the appearance of clinical signs or symptoms. This is the pathological state resulting from a disproportion of essential nutrients with or without the absolute deficiency of any nutrient as determined by the requirement of a balanced diet. If a particular nutrients is high in the body beyond the Recommended Dietary Allowance (RDA) for prolonged period then it's level can become toxic and can lead to serious diseases for example vitamin A, D and Iron toxicity in the body has been observed.

4.4 Understanding Nutritional Status

Nutritional status refers to the condition of the health of an individual as influenced by the utilization of the nutrients from food by the body. It can be determined by correlation of information obtained through medical and dietary history, thorough physical examination and laboratory investigation. Nutritional assessment helps in identifying:

   a) Under Nutrition

   b) Over Nutrition

c) Nutritional deficiencies

d) Individuals at the risk of developing malnutrition

e) Individuals at the risk of developing nutritional related diseases

f) The resources available to assist them to overcome nutritional problems.

The nutritional status can be assessed by the following methods:

I. Direct Methods

a) Nutritional Anthropometry

b) Clinical Examination

c) Biochemical tests and

d) Biophysical methods.

II. Indirect Methods

a) Vital statistics of the community

b) Assessment of socio – economic status and

c) Diet surveys

The basic measurements which should be made on all age groups are weight in kg, length/height and arm circumference in cms. In young children it should be supplemented by measurements of head and chest circumference.

The information on food and nutrient consumption is compared with the recommended allowances given by ICMR and the adequacy is determined. A combination of dietary, clinical and biochemical assessment is desirable for assessment of nutrition status of individuals or communities.

Questions

Q. 1. Define nutrition, nutritional status and health.

Q. 2. Explain what is malnutrition?

Q. 3. How is health and food related?

References

1. Dietetics by Srilakshmi. B, 7th Edition, New Age Publisher, ISBN : 978-81-224-3500-9.

2. Food Science by Srilakshmi. B, 5th Edition, New Age Publisher, ISBN : 978-81-224-2724-0.

3. Textbook on Nutrition and Dietetics for Higher Secondary, Text Book Corporation, College Road, Chennai. Government of Tamil Nadu, First Edition – 2004

4. Jelliffe, D.B., 1989, The Assessment of Nutritional Status of the Community WHO Monograph Series, Geneva

5.   Rotimi E. Aluko, Functional Foods and Nutraceuticals, Springer, ISBN 978-1-4614-3479-5

6.   2010 revised Recommended Dietary Allowances suggested by Indian Council of Medical Research.

7.   Food Lipids: Chemistry, Nutrition and Biotechnology, Antioxidant Mechanisms by Eric A. Decker, Chapter 18

8.   www.eatright.org

9.   http://www.who.int/nutrition/publications/bmi_asia_strategies.pdf

2

Balanced Diet

2.0 Introduction to Balanced Diet

A balanced diet is one which provides all the nutrients in adequate amounts and in proper proportions. This can easily be achieved by properly and judiciously including foods from all food groups. However, the quantities of foods required to meet the nutrient requirements will vary with age, gender, physiological status and physical activity status. In general a balanced diet should provide around 50-60 per cent of total calories from carbohydrates, preferably from complex carbohydrates, about 10-15 per cent from proteins and 20-30 per cent from both visible and invisible fat.

In addition, a balanced diet should provide other non-nutrients such as dietary fibre, antioxidants (vitamins C and E, beta-carotene, riboflavin and selenium) and phytochemicals (polyphenols, flavones etc.) which have positive health benefits. Antioxidants such as protect the human body from free radical damage. Phytochemicals such as polyphenols, flavones, etc., also protect against oxidant damage. Spices like turmeric, ginger, garlic, cumin and cloves are rich in antioxidants.

A balanced diet includes foods from every food group in the recommended amounts. The three considerations for planning balance diet are- Variety (eating many different foods from each food group), Moderation (means keeping portions sizes under control) and Proportionality (how much food a person should choose from each group). Conventionally foods are grouped as:

1. Cereals and millets
2. Pulses and legumes
3. Vegetables and fruits

4. Milk, egg, meat and fish and their products

5. Oils and fats and nuts and oilseeds

However, foods may also be classified according to their functions as discussed in section 2.1. Our diet must provide adequate calories, proteins and micronutrients to achieve maximum growth potential. Therefore, it is important to have appropriate diet during different stages of one's life. There may be situations where adequate amounts of nutrients may not be available through diet alone. In such high risk situations where specific nutrients are lacking, fortification of foods with the limiting nutrient(s) become necessary for example salt is fortified with iodine and iron. Food plays very important role during different stages of life as tabulated below in Table 2.1.

Table 2.1: Importance of Diet during different Stages of Life

Stage of Life	Purpose of Food
Old Adult	For being healthy and physically active Nutrient- dense low fat foods should be considered for this group.
Young Adult	For maintaining health, productivity and prevention of diet-related disease and to support pregnancy/lactation.Nutritionally adequate food with extra nutrient for child bearing/rearing should be considered for this group.
Adolescence	For growth, development, maturation and bone development.Body building and protective foods should be included more in this group.
Young Children's	For growth, development and to fight infections.Energy-rich, body building and protective foods (milk, vegetables and fruits) should be considered for this group.
Infant	For appropriate growth and to protect against infection.Breast milk, energy-rich foods (fats, sugar) should be considered for this group.

How One can make a Balanced Diet?

1. Choose a variety of foods in amounts appropriate for age, gender, physiological status and physical activity. Match food intake with physical activity (sedentary, moderate or heavy worker).

2. Use a combination of whole grains, grams and green vegetables to include dietary diversity.

3. Prefer fresh, locally available vegetables and fruits in plenty.

4. Include in the diets, foods of animal origin such as milk, eggs and meat, particularly for pregnant and lactating women and children.

5. Adults should choose low-fat, protein-rich foods such as lean meat, fish, pulses and low-fat milk.

6. Include jaggery or sugar and cooking oils to bridge the calorie or energy gap.

7. Avoid fried, salty and spicy foods.

8. Consume adequate water to avoid dehydration.

9. Develop healthy eating habits and exercise regularly and try to avoid sedentary lifestyle.

2.1 Functions of Food

Foods are contains different kinds of substances known as the "nutrients." Which when consumed in adequate amounts, fulfil all the functions of the body. Broadly there are six general classes or kinds of nutrients found in all foods namely carbohydrates, fats, proteins, vitamins, minerals and water. Food performs many functions; it has role- physiological functions of the body, social function and psychological function. Each of these functions are discussed below:

2.1.1 Physiological Functions

Each nutrient has its own physiological functions in the body. So, the foods can be classified according to their functions in the body. Three food group classification based on physiological functions of food.

I. Energy Yielding

This group of foods includes food rich in carbohydrate, fat and protein. As, One gram of fat gives 9 calories, one gram of carbohydrate gives 4 calories and one gram of protein gives 4 calories. This group may be broadly divided into two groups:

a. Cereals, pulses, nuts and oilseeds, roots and tubers example wheat, tapioca, tubers, almond, legumes etc.

b. Pure carbohydrates like sugars and fats and oils.

Cereals provide in addition to energy large amounts of proteins, minerals and vitamins in the diet. Pulses also provide protein and B vitamins besides giving energy to the body. Nuts and oilseeds are rich in energy yielding as they are good sources of fats and proteins. Roots and tubers though mainly provide energy, they also contribute to some extent to minerals and vitamins.

Pure carbohydrates like sugars provide only energy (empty calories) and fats provide concentrated source of energy and fat soluble vitamins.

Carbohydrates make up the bulk of our diet. They are our chief source of energy. About 70 percent of the energy (glucose) requirements for all body functions is obtained from carbohydrates. Energy is produced by the oxidation or "internal burning" (cellular respiration) of carbohydrates in the animal cells using oxygen. Carbohydrates also help in the utilization of proteins and fats. When consumed in excess carbohydrates are converted into fats and some amount to glycogen also these serve as energy reserves to be used when required. The main sources of carbohydrates in the diet are starch and sugar. The sources of the former are mainly cereal grains (wheat, rice, millets etc.) or tubers (potato, sweet potato, cassava) and those of the latter are sugarcane and fruits.

Fats or lipids are known as most concentrated form of energy in the food as they furnish more than twice the number of calories per gram as provided by carbohydrates or proteins. When compared to carbohydrates, fats contain a less percentage of oxygen and more of hydrogen, and consequently on oxidation yield more energy. Generally about 30 per cent of human energy requirements are met by fats. When excess energy is supplied to the body, it is stored as fat. Fats are abundant in both plant and animal food. In plants they may be confined to the cytoplasmic membrane or may also be present as reserve material. Fat up to about 15 per cent is present in the germ of cereals. Nuts, such as groundnuts, are rich sources of fats. Butter from milk is an important source of fat. The adipose tissue of animals consists mainly of fats. Fruits are poor source of fat or lipid (except avocado and olive).

II. Body Building

Foods rich in protein are known as body-building foods. These foods are classified into two groups:

   a. Milk, egg, meat, fish: They are rich in proteins of high biological value. These proteins have all the essential amino acids in correct proportion for the synthesis of body tissues.
   b. Pulses, legumes, oilseeds and nuts: They are rich in protein but may not contain all the essential amino acids required by the human body.

Proteins are the major source of building material for the body. They play an important role as structural constituents of cellular membranes and function in the maintenance and repair of body tissues. Proteins also function as biocatalysts. The food value of the protein depends upon the nature and content of its amino acids. When the energy supply to body is not sufficient then the body is forced to use protein as a source of energy by the process of gluconeogenesis. Proteins are found in both animal and plant tissues. Meat, fish, poultry, eggs, milk and cheese are good sources of protein foods from animal sources. Pulses and cereals contain considerable amounts of storage proteins. Soya bean contains over 40 per cent protein on dry weight basis. Nuts and seeds are also good sources of proteins. Starchy vegetables are just fair source of protein as some contain up to 2 per cent protein. Other vegetables and fruits are poor sources of proteins.

III. Protection and Regulation

Foods rich in protein, vitamins and minerals perform regulatory functions in the body for example regulation of hormones, regulation of activities such as maintenance of body temperature, maintaining the heart beat, muscle contraction, control of water balance, blood clotting, removal of waste products from the body and to improve body's immune system or resistance to diseases. Protective foods are broadly classified into two groups.

   a. Foods rich in vitamins, minerals and proteins of high biological value for example milk, egg, fish, liver.
   b. Foods rich in certain vitamins and minerals only for example green leafy vegetables and fruits.

Vitamins are known as "accessory nutrients" as they are required for the proper utilization of the food or nutrients of the diet - carbohydrates, fats and proteins and for the maintenance of good health. Vitamins along with minerals are involved in small quantities in the regulation of body processes. They are constituents of enzymes, which function as catalysts for many biological reactions within the body. Vitamins are found in plant and animal tissues. Their content in plant tissues varies widely depending upon the growing condition, stage of maturity, handling, processing and storage of these food materials. They are not uniformly distributed in plant tissues. Fruits and vegetables are good sources of vitamins. Wheat is an excellent source of B-vitamins, but same is lost with the milling or processing as the bran and germ contain the major proportion of these nutrients.

Minerals also act as catalysts for many biological reactions within the body. Their other functions include the building of bones and other structural parts of the body, muscular contraction, transmission of messages through the nervous system, and the digestion and utilization of nutrients in food. Some minerals like calcium, phosphorus, iron, magnesium and sulphur are required in large quantities, while others like zinc, copper, iodine, manganese, cobalt, etc., are required in small quantities. Minerals are found in foods from animal and plant sources. The mineral content of plant foods varies depending upon the mineral elements present in the medium of cultivation. The distribution of a particular mineral element varies in different tissues.

2.1.2 Social Functions

The social function of food refers to the central part of our social existence. It has been a part our community, social, cultural and religious life. Special foods are distributed as a PRASAD in the religious functions at homes and temples. Feasts are given at specific stages of life such as birth, marriages, birthdays and even death. As food is an integral part of our social existence. This function of food is important in daily life. Food is also a symbol of hospitality and friendship throughout the world. We express our hospitality to a guest by offering for food or a drink. In our country offering a cup of coffee or tea is a symbol of friendship. In general at our homes we serve certain foods for family meals and different foods when we have guests. In some parts of the world it is prestigious to use polished white rice instead of brown rice (which protects one against beriberi). The same is true of white flour and white bread which have replaced "brown or multigrain bread."

Food is a status symbol. The status factor is also associated with certain foods used by the so-called upper class families. Social function of food is also seed in the times of disaster or sorrow as we take food to the affected persons, for example we carry fruits along with us when we go to meet relatives or friends in joy and even when relative is ill.

2.1.3 Psychological Functions

In addition to satisfying physiological and social needs. Food must satisfy certain emotional need as well. This includes a sense of security, love and attention. Thus, eating or just listening about familiar foods make us feel secure. These sentiments are the bases of normal attachment to the mother's cooking. These aspects are important in food acceptance and must be considered in planning meals, which are not only nutritionally adequate but also enjoyable for the group for whom they are intended. Food is an outlet for emotion. As a relief from tension one may eat or overeat. For some people loneliness and boredom are relieved by continuous nibbling of food. Anger and frustration may turn one against food. Specific foods are associated with unhappy experiences. Foods consumed by some people are unacceptable and even revolting for others. Such concepts have no rhyme or reason, nor are they related to nutritive value; they are just emotional reactions. Thus, for an average man, food is much more than a substance supplying nutrients for health. It is the sum of his culture and traditions, emotional outlet, gratification of pleasure, and a relief from stress, a means of communication, security, status— all of these interwoven in the fabric of life and unconsciously expressed in food likes and dislikes. The psychological and emotional reactions to food do not yield easily to reasoning or scientific facts about nutrition.

Food is a source of security. An infant feels security by the way his mother feeds him and develops a sense of security with mother. Even individual's behavioural patterns will be influenced by the extent to which they feels secure with regards his food supply. Similarly a growing child gains confidence and a feeling of belonging when he knows there is food in the house and he will be fed. People feel reasonably secure when they have enough food stored up to take care of them during periods of scarcity. Familiar foods give a sense of security when one has to eat away from home.

The food may be used in society and in the family as a way to reward or punishment. A child may be sent to bed without food because of bad behaviour or may be given something special as a reward for some achievement. Denial of food is often being used as a revolt against "the establishment or government."

Cultural knowledge, beliefs, religion, and traditions teach individuals which foods should be considered proper or appropriate to eat and which are not. For example, hindu people don't eat cow meat as it is considered a sin but at the same time other for some other community it is proper to eat same. In some part of the world people eat foods such as blood, mice, snake and insects etc. even though these foods are packed with nutrients and safe to eat, but at the same time people of other parts of the world don't feel they are proper to eat. Some early experiences with people, places, and situations influence lifelong food choices. It can also be concluded that our diet patterns begin when our parents introduce us to foods as children.

2.2 Food Groups

Foods have been classified into different groups depending upon the nutritive value, for the convenience of planning diets. In planning balanced diets, food should be chosen from each group in sufficient quantity. Cereals and pulses should be taken adequately; fruits and vegetables liberally; animal foods moderately and oils and sugars sparingly. Food groups like 'Basic four', 'Basic five' or 'Basic seven' can be used for planning diets as per the convenience.

I. Basic Four

Group	Nutrient
Cereals, millets and pulses	Energy, protein and B-vitamins
Vegetables and fruits	Vitamins, minerals and fibre
Milk, milk products, and animal foods	Protein, calcium and B-vitamins
Oils, fats, nuts and oilseeds	Energy and protein (nuts and oilseeds).

II. Basic Five: ICMR

Group	Nutrient
Cereals, grains and products: rice, wheat, ragi, maize, bajra, jawar, rice flakes, puffed rice.	Energy, protein, invisible fat, thiamine, folic acid, riboflavin, iron and fibre.
Pulses and legumes: Bengal gram, black gram, cow pea, peas (dry) rajma, soyabeans.	Energy, protein, invisible fat, thiamin, riboflavin, folic acid, calcium, iron and fibre.
Milk and meat products: – Milk, curd, skimmed milk, cheese – Chicken, liver, fish, egg and meat.	 – Protein, fat, riboflavin – Calcium, protein, fat, riboflavin.
Fruits and vegetables: – Mango, guava, tomato, papaya, orange, sweet lime, watermelon. – Green leafy vegetables: Amaranth spinach, gogu, drumstick leaves, corriander leaves, fenugreek. – Other vegetables: Carrots, onion, brinjal, ladies finger, beans, capsicum, cauliflower, drumstick.	 – Carotenoids, vitamin C, riboflavin, folic acid, iron, fibre. – Riboflavin, folic acid calcium, fibre, iron, carotenoids. – Carotenoids, folic acid, calcium and fibre.
Fats and Sugars: – Fats: Butter, ghee, hydrogenated fat, cooking oils. – Sugar and jaggery.	 – Energy, essential fatty acids and fat soluble vitamins. – Energy (Jaggery also contain iron).

III: Basic Seven

Group	Nutrient
Green and yellow vegetables	Carotenoids, ascorbic acid, and iron.
Oranges, grape fruit, tomatoes, raw cabbage	Ascorbic acid.
Potatoes, other vegetables and fruits	Vitamins and minerals in general and fibre.
Milk and milk products	Calcium, phosphorus, protein and vitamins
Meat, poultry, fish and eggs	Proteins, phosphorus, iron and B vitamins.
Bread, flour and cereals	Thiamine, niacin, riboflavin, iron, carbohydrate and fibre.
Butter or fortified margarine	Vitamin A and fat

2.4 Food Guide Pyramid

Figure 2.1 shows the food guide pyramid. It is meant for use by the general healthy population as a guide for the amount and types of foods to be included in the daily diet. Earlier in United State's USDA has laid a food pyramid as given in figure 2.2 to help the consumer to eat balanced diet. But now the concept has moved from food pyramid to "My Pyramid" concept which is more personalized.

The goal of MyPyramid, Steps to a Healthier You, is to provide advice that helps consumers can adopt to live longer, better and have healthier lives. MyPyramid was designed to depict elements that form the basis of a healthy diet and lifestyle: variety, proportionality, moderation, personalization, physical activity, and gradual

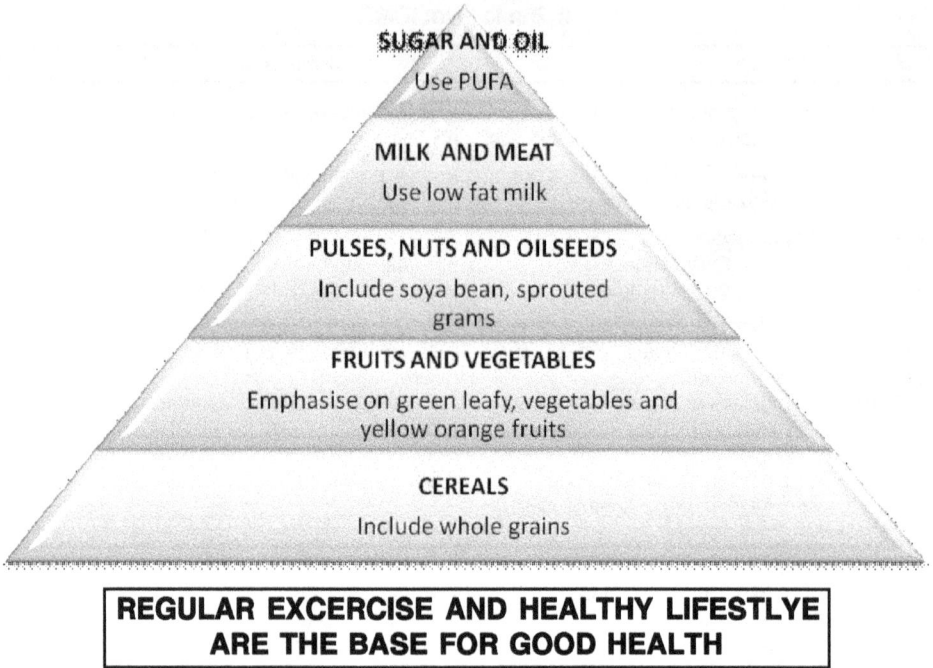

Figure 2.1: ICMR Food Pyramid.

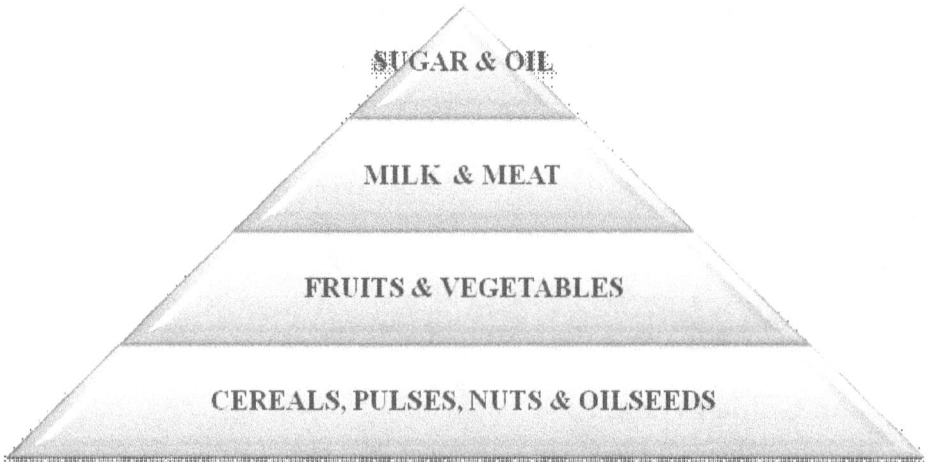

Figure 2.2: USDA Food Pyramid.

improvement. The nutritionally adequate or balanced diet can be planned using MyPyramid by selecting a variety of foods, including grains, vegetables, fruits, milk, meat and beans, and oils.

There is MyTracker program at the www.MyPyramid.gov website which can help consumer to monitor progress. My Pyramid is presented in Figure 2.3 and the various components of MyPramid are :

| Legends: | Grains | Vegetables | Fruits | Oils | Milk | Meat & Beans |

Figure 2.3: My Pyramid.

1. **Activity:** Activity is represented by the steps and the person climbing them, as a reminder of the importance of daily physical activity.
2. **Moderation:** Moderation is represented by the narrowing of each food group from bottom to top. The wider base stands for foods with little or no solid fats or added sugars. These should be selected more often. The narrower top area stands for foods containing more added sugars and solid fats.
3. **Personalization:** Personalization is shown by the person on the steps.
4. **Variety:** Variety is represented by the 6 colour bands representing the 5 food groups of the food pyramid. This exemplify that foods from all food groups is needed each day for maintain good health.
5. **Proportionality:** Proportionality means in what proportion you should eat from each food group, this is shown by the different widths of the food group bands. However the widths are just a general guide, not exact proportions.

Questions

Q-1. What is a balanced diet and what are its components?

Q-2. How is food classified in a three food group system?

Q-3. What is Psychological function of food?

Q-4. How a food pyramid different from My Pyramid?

References

1.   Dietary guidelines for Indians - a Manual, national institute of nutrition Hyderabad – 500 007, India, Second Edition, 2011.

2.   Food Science by Srilakshmi. B, 5th Edition, New Age Publisher.

3.   Shakuntala Manay, M. and Shadaksharaswamy, M., (1987), Foods-Facts and Principles, New Age International (P) Publishers Ltd., Chennai.

4.   Wardlaw's, Perspectives in Nutrition, 8th Edition, McGraw-Hill Companies.

5.   www.MyPyramid.gov.

Unit II: Nutrients

3
Energy

3.0 Introduction

Energy can be defined as the ability to work. Energy is essential for rest, growth and other activities. Even during resting; energy is required for respiration, blood circulation, digestion, absorption, excretion and even for maintaining bodily temperature etc. The energy to perform work is derived from the carbohydrate, fat and protein in the diet (see subsection 3.3.1). The source of energy in diets varies depending on farming practices, cultural, social and economic factors. The body needs energy for maintaining body temperature, metabolic activity, supporting growth, for physical work, to maintain constant body weight and good health. The body's storage energy or potential energy is continuously available in the body from the glycogen in muscle and liver. This stored energy is transformed to other forms to accomplish the work of the body. Examples are

    I. Osmotic Energy – Maintain transport of nutrients.

    II. Electrical Energy – Transmission of nerve impulse.

    III. Chemical Energy – Synthesis of new compounds.

    IV. Thermal Energy – Heat regulation.

Whenever one form of energy is produced, another form is reduced by exactly the same amount as stated by the Law of Conservation of Energy. This law states that energy can neither be created or destroyed it can only be transformed from one form to another.

3.1 Units of Energy – Calorie and Joule

The unit of energy, kilocalorie (kcal) was used for a long time. Recently the International Union of Sciences and International Union of Nutritional Science (IUNS) have adopted 'Joule' as the unit of energy in place of kcal. These units are defined as follows.

☆ A joule is defined as the energy required to move 1 kg mass by 1 meter by a force of 1 Newton acting on it.

☆ One Newton is the force needed to accelerate 1 kg mass by less than a second.

☆ Kcal is defined as the heat required to raise the temperature of 1kg of water by 1° C. (From 16.5°C to 17.5°C).

1 Kcal   = 4.184 KJ (Kilo Joules)

1000 Kcal = 4184 = 4.18 MJ (mega joules)

1 KJ   = 0.239 Kcal.

3.2 Energy Value of Foods

The energy in various foods is measured by calorimetry. Calorimetry is the measurement of heat loss. An instrument for measuring heat output of the body or the energy value of foods is called a Calorimeter. Bomb calorimeter is used for measuring the calorie value of foods.

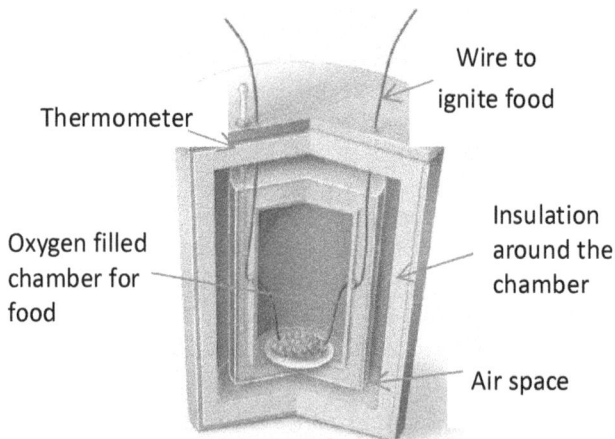

Figure 3.1: A Pictorial Depiction of Bomb Calorimeter.
***Source*: Perspectives in Nutrition, 8ᵗʰ edition, Gordon M. Wardlaw, Pal M. Insel.**

Bomb calorimeters measure calorie content by igniting and burning a dried portion of food. The burning food raises the temperature of the water surrounding the chamber holding the food. The increase in water temperature indicates the number of kilocalories in the food (because 1 kilocalorie equals the amount of heat needed to raise the temperature of 1 kg of water by 1°C). The maximum amount of energy that the sample is capable of yielding when it is completely burnt or oxidized is the energy value of that food, also known as heat of combustion. The energy measured using a Bomb Calorimeter is as follows

☆ 1 g of Carbohydrate - 4.1 kcal

☆ 1 g of fat - 9.45 kcal

☆ 1 g of protein - 5.65 kcal

When samples of carbohydrate, fat, protein are burned, the amount of heat produced is always the same for each of these nutrients. In the bomb calorimeter carbohydrates, fats and proteins are completely oxidized whereas in the human body the process of digestion and absorption does not proceed with 100 percent efficiency. The extent of digestion varies from one nutrient to another. The Coefficient of digestibility is used to express the proportion of an ingested nutrient that ultimately becomes available to the body cells. The coefficient of digestibility for carbohydrate, fat and protein are 0.98, 0.95 and 0.92 respectively.

It is observed that carbohydrate and fat are metabolized almost completely, whereas protein metabolism is incomplete due to the presence of nitrogen. The physiological energy value of carbohydrate, fat and protein are 4, 9 and 4. These values are known as Atwater Bryant factors or physiological fuel values as given in Table 3.1.

Table 3.1: Physiological Fuel Value of Carbohydrate, Fat, Protein

	Heat of Combustion kcal	Coefficient of Digestibility	Digestibility Per cent	Physiological Fuel Value kcal
Carbohydrate	4.1	0.98	98	4.0
Fat	9.45	0.95	95	9.0
Protein	5.65	0.92	92	4.0

Source. Robinson C. H., Marilyn R. and Lawler. 1982. Normal and Therapeutic Nutrition.

3.3. Energy Balance

Energy balance is the relationship between energy intake and energy expenditure. To maintain daily energy balance; the total energy requirement of a being is the number of kcal required to restore daily basal metabolic loss and loss from exercise and other physical activities.

Energy equilibrium occurs when the calories consumed from food source (energy intake) is equal to amount of energy expended.

Positive energy balance results if energy intake exceeds energy expended. The excess energy consumed is stored, resulting in weight gain. Positive energy balance is desired during the growth stages of the life cycle (pregnancy, infancy, childhood, adolescence) and to restore body weight to healthy levels after losses caused by starvation, disease, or injury. However, during other times, such as adulthood, positive energy balance over time can cause body weight to climb to unhealthy levels. The process of aging itself does not cause weight gain; rather, weight gain stems from a pattern of excess food intake coupled with limited physical activity and slower metabolism.

Negative energy balance results when energy intake is less than energy expenditure. Weight loss occurs because energy stored in the body in the form of fat or glycogen in liver and muscle is used to make up for the shortfall in energy intake. Negative energy balance is desired in adults when body fatness exceeds healthy levels. Negative energy balance during growth stages of the life cycle generally is not recommended because it can impair normal growth.

Components of Energy Balance

3.3.1 Energy Intake

Sources of Energy

Food as Fuel for Body: Energy is required for whatever work a human does whether physical or mental. Humans both consume and spend energy which is expressed in terms of calories. The calories from food are derived from carbohydrates, lipids, and proteins.

Food is a Source of Following Macronutrients from which the Energy is Derived for Body

Carbohydrate	Lipids	Proteins
– Main energy nutrient, provides around 4calories/g of energy. – In the process of digestion the large carbohydrate molecules are broken down into monosaccharide's, primarily glucose. – The glucose so produces is absorbed into the body and is then circulated through the blood supply to replenish blood glucose level and then any excess glucose is then stored as glycogen in liver and muscle for future long term energy requirements.	– Fat is a remarkable fuel for body as it provides around 9 calories/g, this amount is twice the amount of calories provided by carbohydrate or protein. – Lipids are triglycerides and are made up of three fatty acids attached to a glycerol molecule. During the process of digestion or energy requirement, fatty acids are broken down and are released into the bloodstream. – All body cells use fat for fuel except the brain, nervous tissues and red blood cells which require glucose for their functioning.	– Proteins provides around 4calories/g of energy. – Although the prime function of proteins is to give structural support to body cells and tissues but when sufficient glucose is not available or the energy needs are higher then body will start converting its protein reserve into energy. – During the process when proteins are used as body fuel, nitrogen containing amine group will be used, few of the amino acids will be transformed to glucose, which can be used by tissues which cannot use fatty acids. But in either of ways, reconversion back into proteins is not possible. – If excess proteins is consumed it will be converted into fat.

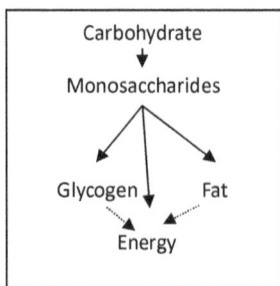

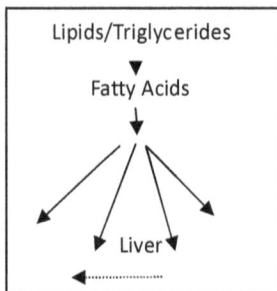

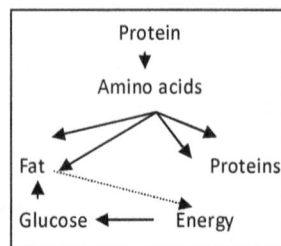

Carbohydrate → Monosaccharides → Glycogen, Fat → Energy

Lipids/Triglycerides → Fatty Acids → Liver

Protein → Amino acids → Fat, Proteins; Glucose ← Energy

The amount of energy in a food can be derived by directly measuring calorie content using a device called a bomb calorimeter (Figure 3.1). Calorie content is most commonly calculated by determining the grams of carbohydrate, protein, and fat (and possibly alcohol) in a food and multiplying these compounds by their physiological fuel values. (refer 3.2 subsection above, the physiological fuel values are 4 kcal/g for carbohydrates and proteins, 9 kcal/g for fat, and 7 kcal/g for alcohol.)

What are the Factors which Affect Energy Input?

1. Appetite

This refers to nonphysiological or external regulation of eating. This include physiological, environmental factors, food accessibility, and food characteristics.

2. Hunger

This refers to internal regulation for eating. This include, stomach distension or satiety, physical activity.

3. Food Liking

This refers to the choice of people what they choose to eat. For example people who like fried food or food with high amount of fat are more likely to receive more calories then a meal rich in fiber.

3.3.2 Energy Output or Expenditure

The body uses energy for 3 universal purposes:

1. Basal metabolism;
2. Physical activity;
3. For the digestion, absorption, and processing of ingested nutrients.
4. Thermogenesis (minor form of energy output)

   Above mentioned inputs and outputs are depicted in Figure 3.2.

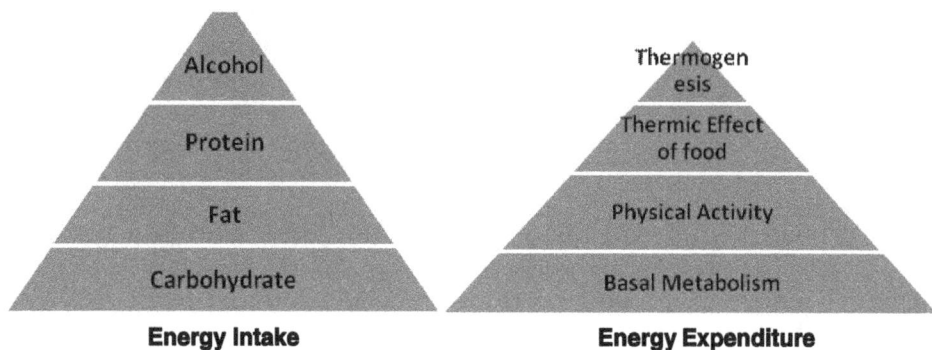

Figure 3.2: Major Components of Energy Intake and Expenditure. The size of each component shows the relative contribution of that component to energy balance.

* Alcohol is an additional source of energy only for those who consume it.

3.3.2.1 Basal Metabolism

Basal Metabolism is the minimum amount of energy needed by the body for maintenance of life when the person is at post absorptive state, physical and emotional rest. Thus Basal Metabolism represented as basal metabolic rate (BMR) represents the minimum amount of energy expended in a fasting state (12 hours or more) to keep a resting, awake body alive in a warm, quiet environment. For a sedentary person, basal metabolism accounts for about 60 to 70 per cent of total energy expenditure. Some of the processes involved include the beating of the heart, respiration by the lungs, and the activity of other organs, such as the liver, brain, and kidney. It does not include energy expended for physical activity or the digestion, absorption, and processing of nutrients recently consumed.

If the person is not fasting or completely rested, the term resting metabolism is used (expressed as resting metabolic rate [RMR]). RMR is typically 6 per cent higher than BMR. Both BMR and RMR are expressed as the number of calories burned per unit of time. Energy is required even when an individual is resting. Table 3.2 lists the approximate percentages of total energy used by the body's organs at rest.

Table 3.2: Energy Required/Day for Functioning of Various Body Organs at Rest

Organ	Energy/day (kcal/day)	Per cent of Energy
Liver	380 kcal/day	27 per cent
Brain	265 kcal/day	19 per cent
Skeletal muscle	250 kcal/day	18 per cent
Kidney	140 kcal/day	10 per cent
Heart	100 kcal/day	7 per cent
Other	265 kcal/day	19 per cent

Source: Perspectives in Nutrition, 8th edition, Gordon M. Wardlaw, Pal M. Insel.

The basal metabolic rate can be measured directly from the heat produced (using a Respiration Calorimeter and Metabolic Chamber) or indirectly from O_2 intake and CO_2 expenditure when the subject is at rest. Physical activity including manual work, which can be sedentary, moderate or heavy work requires additional supply of energy. Even in absence of any physical work, individuals have to do some activities like walking, sitting, studying, climbing stairs etc. this is called as "**maintenance energy**" for which roughly 1.5 times of basal energy is required. The required energy for both is derived from dietary sources. A rough estimate of basal metabolic rate for women is 0.9 kcal/kg per hour and 1.0 kcal/kg per hour for men. Thus, BMR accounts for the largest number of the calories expended by an individual each day.

How to Calculate BMR?

1. One way is to calculate BMR by using the BMR factor. For example for a female weighing 60kg, with average BMR for female @0.9kcal/kg per hour, her BMR will be:

   BMR/hour = 60 kg× 0.9kcal/kg per hour = 54 kcal/hour

   BMR/day = 54 × 24 = 1296 kcal/day

2. Calculation of energy requirements by Predicting BMR:

Equations for predicting BMR (kcal/24 hrs) as proposed by the ICMR expert committee for Indians is given in Table 3.3.

Table 3.3: Equations for Predicting BMR (kcal/24 hr)

Gender	Age (yrs)	Prediction Equation
Male	18 – 30	14.5 x B.W (kg) + 645
	30 – 60	10.9 x B.W (kg) + 833
	> 60	12.8 x B.W (kg) + 463
Female	18 – 30	14.0 x B.W (kg) + 471
	30 – 60	8.3 x B.W (kg) + 788
	> 60	10.0 x B.W (kg) + 565

Source: ICMR 2002 Nutrient requirements and recommended dietary allowances for Indians, NIN.

Various factors affecting basal metabolic rate as are listed in Table 3.4.

Table 3.3: Factors Affecting Basal Metabolic Rate (BMR)

Sl.No.	Factor	Effect on BMR
1.	Body Composition	1. The more lean body mass, higher is the BMR. This is due to greater metabolic activity in these tissues when compared to bones and fat. Men with a high proportion of muscle mass or lean body mass have a higher BMR than women. 2. Larger is the body surface area more will be the basal metabolism.
2.	Gender	Male gender typically have more lean body mass than females so, Men will have more BMR compared with Females.
3.	Body temperature (fever or cold environmental conditions)	Fever increases the BMR. There is a 7 per cent increase in BMR for each degree rise in temperature in Fahrenheit.
4.	Nervous system activity like Stress	Stress raises BMR.
5.	Tobacco (Smoking) and Caffeine	Increases the BMR
6.	Hyperthyroidism (Over secretion of thyroxin-thyroid hormone)	The basal metabolic rate is increased as much as 50-70 per cent.
7.	Growth	In children and pregnant women the BMR is higher.
8.	Pregnancy	During the last trimester of pregnancy basal metabolic rate increases by 15- 25 per cent as there is a increase in muscle mass of uterus, size of mammary gland, foetal mass and placenta, cardiac work and respiratory rate.
9.	Fasting/Starvation	Lowers BMR
10.	Hypothyroidism (under secretion of thyroxin)	The basal metabolic rate is decreased by 30 per cent.
11.	Age	Lean body mass diminishes with age slowing the BMR.
12.	Height	In tall people the BMR is higher.
13.	Undernutrition	Prolonged undernutrition lowers the BMR.
14.	Food intake and Recent exercise	Recent food intake and recently done exercise (effect has limited duration) increases the BMR.

Basal metabolism decreases by about 10 to 20 per cent (about 150 to 300 kcal/day) when calorie intake declines and the body shifts into a conservation mode. This shift helps us to survive during periods of famine and starvation, but it also is a barrier to sustained weight loss during dieting that involves extremely low calorie intake. Basal metabolism drops 1 to 2 per cent for each decade past the age of 30 years as a result of the lean body mass loss that typically occurs with advancing age. However, physical activity helps maintain lean body mass and helps preserve BMR throughout adulthood.

3.3.2.2 Physical Activities

Physical activity, contribute to the largest energy expenditure even above and beyond basal energy requirement by as much as 25 to 40 per cent. Individuals vary in their physical activity based on their profession or occupational activity and individual preference like walking instead of driving to a nearby grocery store or climbing stairs instead of opting for escalator etc., hence leading to energy expenditure. Our inactivity to do any type of physical activity is a major cause of increasing rate of obesity in many countries.

An individual should do atleast Forty five minutes per day of moderate intensity physical activity to get health benefits. However, even better health benefits can be achieved through more vigorous workout or by staying active for a longer duration. This will burn more calories. An individual can perform all physical activity either at once or divide it into two or three parts during the day.

For calculating energy requirements, the professions have been classified as sedentary, moderate and heavy. The energy cost of physical activities is expressed in terms of BMR units. The ICMR classification of activities based on occupation is given in Table 3.5.

Table 3.5: Classification of Activities Based on Occupation

Gender	Sedentary (80-180 kcal/hr)	Moderate (170-240 kcal/hr)	Heavy (250-350 kcal/hr)
Male	1) Teacher 2) Tailor 3) Executive	1) Fisherman 2) Potter	1) Stone Cutter 2) Mine Worker 3) Wood Cutter
Female	1) Teacher 2) Executive 3) Nurse	1) Servant maid 2) Weaver	1) Stone Cutter

Source: ICMR 2012 Nutrient requirements and recommended dietary allowances for Indians, NIN.

BMR Unit

The energy cost of rest and physical activity is expressed as multiples of BMR which is called the physical activity ratio (PAR). The physical activity ratio expresses the energy cost of an individual activity per minute as ratio of the cost of BMR per minute. Hence it is advantageous to express the energy expenditure in terms of BMR units (Table 3.6).

Table 3.6: Energy Cost of some Common Activities in Terms of BMR Units

Activity	Energy Cost of Activities in BMR Units
1. Sitting quietly	1.2
2. Standing quietly	1.4
3. Sitting at desk	1.3
4. Walking (3MPH)	3.7

Source: ICMR 2002 Nutrient requirements and recommended dietary allowances for Indians, NIN.

Using factorial method the WHO/FAO expert committee has derived the BMR factors for Indian men and women as 1.6, 1.9 and 2.5 respectively for the three categories of activities namely, sedentary, moderate and heavy as given in table-3.6.

Table 3.7: Energy Requirements of Indian Adults in Terms of BMR Units

Activity	Duration (hrs)	Rate of Energy Expenditure in Terms of BMR Units		
		Sedentary	Moderate	Heavy
Sleep	8	1.0	1.0	1.0
Occupational activity	8	1.7	2.8	4.5
Non-Occupational Activity	8	2.2	2	–
Average for 24 hr		1.6	1.9	2.5

Source: ICMR 2012 Nutrient requirements and recommended dietary allowances for Indians, NIN.

Calculating Daily Energy Requirement

The energy requirements are calculated using the computed BMR (refer 3.3.2.1) from body weights and recommended BMR factor for Indians for different levels of physical activity (which is 1.6, 1.9 and 2.5 for sedentary, moderate and heavy activity respectively) are arrived at.

For example: For an Indian adult man 25 yrs of age, weighing 60kg and doing moderate activity, the energy requirement is calculated as follows.

1. BMR = 14.5 x 60 + 645 = 1515 kcal/24 hr
2. Energy requirement= predicted BMR x BMR units for activity=1515 x 1.9 =2878.5kcal/day.

3.3.2.3 Thermic Effect of Food

The thermic effect of food (TEF) is the energy the body uses to digest, absorb, transport, store, and metabolize the nutrients consumed in the diet, this overall effect of food is called dietary thermogenesis or thermic effect of food (formerly called as **specific dynamic action (SDA).** Thermogenesis is also known as diet induced thermogenesis (DIT), thermoregulation and non-exercise activity thermogenesis (NEAT). TEF is the increase in energy expenditure above the RMR that can be measured for several hours after a mean. Thermogenesis, the process of heat production by humans and other organisms, that makes a small contribution to overall energy expenditure (for about 5 to 10 per cent of the energy consumed each day but the total amount varies among individuals). For example if daily energy intake is 2000 kcal, TEF would account for 100 to 200 kcal.

Factors which Affects TEF

1. Food composition: For example, the TEF value for a protein-rich food (20 to 30 per cent of the energy consumed) is higher than that of a high carbohydrate (5 to 10 per cent) or high fat (0 to 3 per cent) food or meal because it takes more energy to metabolize amino acids into fat than to convert glucose into glycogen or transfer absorbed fat into adipose reserve.

2. Meal size: Large meals result in higher TEF values than the same amount of food eaten over many hours.

Adaptive Thermogenesis

Adaptive thermogenesis heat is produced when the body expends energy for non-voluntary physical activity triggered by cold conditions or overeating. Examples of such activities include fidgeting, shivering when cold, maintaining muscle tone, and holding the body up when not lying down.

Certain people are able to resist weight increase from overeating by inducing thermogenesis, whereas others have little thermogenesis.

Brown adipose tissue (found in small amounts in infants and hibernating animals) is a specialized form of fat tissue that participates in thermogenesis. The brown appearance is because of the large number of capillaries it contains. Brown fat has a protein (uncoupling protein) that uses the food we consume to generate heat for the body instead of creating energy in the form of ATP. Therefore brown fat releases energy from energy-yielding nutrients as heat thus contributing to thermogenesis. Adults have very little brown fat and its role in adulthood is not known. Brown fat is responsible for thermoregulation during infancy (when brown fat accounts for around 5 per cent of body weight). Brown fat is used by hibernating animals to generate heat during cold winter duration.

3.4 Diseases Related with Energy Imbalance

3.4.1 Obesity

Obesity represents an energy imbalance resulting from an excess of energy input over energy output (expenditure). This means a condition of having too much body fat, it is usually more than 20 per cent above individual ideal weight. There are several health problems associated with obesity.

☆ It is considered to be a psychological and emotional burden which may have serious impact on individuals.

☆ It is related with increased blood pressure. An obese person is 3 times more at a risk than a normal individual.

☆ Increased risk of diabetes. An obese person is 3 times more at a risk than a normal individual.

☆ It is also associated with high levels of cholesterol in the blood, the risk is almost twice as compared with the non-obese person.

☆ It is associated with increased risk of cancer. Obese males have more risk of developing colon, prostate and rectum cancer whereas females are at risk of developing cancer of gall bladder, ovaries, uterus, breast and bile ducts.

☆ Increased risk of heart disease is also associated with the obesity.

☆ Increased risk of early death.

The health risks are greater for those with body fat more in the upper body part. This can be determined by calculating waist to hip ratio. A high ratio can predict elevated complications from obesity.

Factors which Cause Obesity

1. **Genetic and parental influence:** some people are at more risk of developing obesity that others due to genetic factors. These kinds of people are usually more careful about their eating behavior and exercise regime to counteract their inherited tendency to gain weight.

   Genetic factor is related with fat cell an individual has. According to fat cell theory, if an individual takes in too many calories; fat cell number can increase to two or three times than normal. And also include those which are formed, the extra fat cells cannot be removed by the body. This can happen anytime during the life span of an individual but is particularly significant during infancy, when the fat cells are still dividing. Infants who are over-nourished tend to remain overfat as children because of the extra fat cells formed and will become obese children, teenagers and can even remain obese as adults. One theory also suggests that a mother who breast feeds her baby may help prevent formation of extra fat cells another theory states that if solid food is introduced after nine month this may help prevent infant obesity.

2.
   Dietary factors: This factor means when the individual likes to consume lot of foods especially calorie rich foods like foods high in fat (energy dense foods).

3. **Low energy expenditure:** An individual can loose or burn calories during the entire day span in various activities which includes BMR, resting metabolic state, physical activity (including exercise) and Thermic effect of food. If an individual do gym or lot of physical activity then that individual is less likely to gain weight. Most of the studies show that both obese children and adults are less active than normal weight people.

Points to Ponder

1. Slow and steady reduction in body weight is advisable.
2. Severe fasting may lead to health hazards.
3. Achieve energy balance and appropriate weight for height.
4. Encourage regular physical activity.

How to Treat Obesity?

Obesity	Mild Obesity	Moderate Obesity	Severe Obesity
Meaning	– Refers to the condition of being weighing 20-40% overweight. This is the most common form of obesity. – This condition require medical supervision due to the high amount of body fat.	– Refers to the condition of being weighing 41-100% overweight. – This condition require medical supervision due to the high amount of body fat.	– Refers to the condition of weighing more than double the ideal weight. – This condition is associated with severe medical complications.
Treatment	This involves controlled intake of calories by reducing fat intake, increasing the intake of complex carbohydrate, by increasing energy expenditure by increasing physical activity, by behavioral modification which includes monitoring meals of self either by maintaining diet diaries, giving rewards to self for controlling excessive meal.	This type of obesity involves eating very low calorie diet to provide 400-700 calories per day. However along with this eating protein is emphasized to avoid loss of muscle tissue.	This type of obesity involves the patient to undo surgery (liposuction) and is forced to change its eating habits drastically.

5. Eat small meals regularly at frequent intervals.
6. Cut down sugar, salt, fatty foods, refined foods, soft drinks and alcohol.
7. Eat complex carbohydrates, low glycemic foods and fibre rich diets.
8. Increase consumption of fruits and vegetables, legumes, whole grains and nuts.
9. Limit fat intake and shift from saturated to unsaturated fats.
10. Avoid trans-fatty richfoods (vanaspati, bakery products and sweets).
11. Use low- fat milk.

3.4.2 Underweight

Extreme weight loss is a state of energy imbalance resulting from a deficit energy input over energy output. This could be due to illness, food disliking etc. About 1/4th of the world's children are underweight and almost a third have underdeveloped growth. Such children are more likely to suffer from infectious diseases and to have problems learning. Many females or girls have less or restricted access to food than men or boys in the house because of the social customs which says females or girls should eat last. When a woman is inadequately nourished, her developing fetus or breastfed infant also may suffer from malnutrition.

Being underweight (BMI below 18.5) also carries health hazards like loss of menstrual function, low bone density, complications with pregnancy and surgery, and slow recovery after sickness. In growing children and teenage, underweight can interfere with normal growth and development. Significant underweight also is associated with increased death rates.

Underweight can be caused by overindulgence in physical activity, restricted calorie intake, mental stress or depression and health conditions such as cancer, infectious disease (*e.g.*, tuberculosis), digestive tract disorders (*e.g.*, chronic inammatory bowel disease) etc. Active children and teens who do not take the time to consume enough calories to support their energy requirement of body may become underweight. Genetic background may confer characteristics that promote low body weight, such as a higher metabolic rate, a lean body frame, or both.

Treatment of Underweight

Treatment involves taking sufficient calories to sustain body's energy requirement. Treatment of the disease or illness conditions, by eating food even if you don't feel like eating. This would help to gain weight.

3.4.3 Chronic Energy Deficiency

Refer section 5.7.2

3.5 Recommended Dietary Allowances (RDA)

Table: 3.7 RDA for Energy

Group	Physical Activity Group	Body Weight (kg)	Energy (kcal/day)
Man	Sedentary, Moderate, Heavy work	60	2425, 2875, 3800
Woman	Sedentary, Moderate, Heavy work	50	1875, 2225, 2925
	Pregnant woman	50	+300
	Lactating woman 0-6 months	50	+550
	Lactating woman 6-12 months	50	+400
Infants	0-6 months	5.4	108/kg
Boys	13-15	47.8	2450
Girls	16-18	49.9	2060

Source. RDA, Nutritive Value of Indian Foods, 2012 ed. By C. Gopalan *et al.*

Questions?

Q. 1. Define kcal and BMR.

Q. 2. What is dietary thermogenesis ?

Q. 3. What is Basal Metabolic rate? What are the factors which affect BMR?

Q. 4. Which equipment be used for measuring energy?

References

1.   Textbook on Nutrition and Dietetics for Higher Secondary, Text Book Corporation, College Road, Chennai. Government of Tamil Nadu, First Edition – 2004

2.   Nutrition, David C. Nieman et.al., Wm. C. Brown Publishers, Ed. 1990.

3.   Wardlaw's, Perspectives in Nutrition, 8th Edition, McGraw-Hill Companies, ISBN 978–0–07–296999–3.

4.   Dietary Guidelines For Indians -A Manual, National Institute of Nutrition Hyderabad – 500 007, INDIA, Second Edition, 2011.

5.   2010 revised Recommended Dietary Allowances suggested by Indian Council of Medical Research.

4

Carbohydrates

4.0 Introduction

The most abundant organic molecules in nature. Plant use CO_2 and Water and energy (from Sun) to Produce CHO. These are organic compounds that contain CARBON, HYDROGEN, and OXYGEN in the ratio of 1:2:1. The general formula is $(CH_2O)n$. They provide a significant fraction of the energy in the diet of most organisms and are important source of energy for cells. Carbohydrates are Polyhydroxy aldehydes or ketones, or substances that yield these compounds on hydrolysis. Carbohydrates performs following functions:

☆ Can act as a storage form of energy

☆ Can be structural components of many organisms

☆ Can be cell-membrane components mediating intercellular communication

☆ Can be cell-surface antigens

☆ Can be part of the body's extracellular ground substance

☆ Can be associated with proteins and lipids

☆ Is part of RNA, DNA, and several coenzymes (NAD^+, $NADP^+$, FAD, CoA).

4.1 Carbohydrate Classification and Structure

There are two main classes of carbohydrates: SIMPLE (sugars) and COMPLEX (starches and fiber) carbohydrates. Simple Sugars (sugar is also known as Saccharide) are the building blocks of complex sugars and polysaccharides. Simple sugars are further classified as:

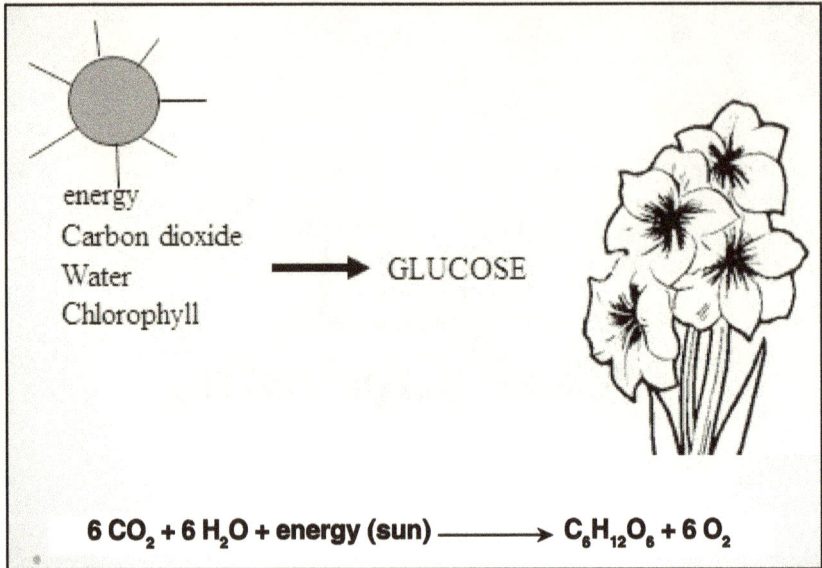

energy
Carbon dioxide
Water
Chlorophyll

GLUCOSE

$$6\ CO_2 + 6\ H_2O + energy\ (sun) \longrightarrow C_6H_{12}O_6 + 6\ O_2$$

Figure 4.1: Photosynthesis: Sun's Energy becomes Part of Glucose Molecule.

Monosaccharides	Disaccharides
☆ Glucose (principal monosaccharide in the body),	☆ Sucrose= Glucose+ Fructose
☆ Fructose	☆ Lactose= Glucose + Galactose
☆ Galactose	☆ Maltose= Glucose + Glucose

4.1.1 Simple Sugar

4.1.1.1. Monosaccharides

These are simple sugars containing short chains of carbon atoms with one aldehydic or ketonic group (carbonyl group), each of remaining carbons bear a hydroxyl group. Polyhydroxy aldehydes or ketones that can't easily be further hydrolyzed are "Simple sugars". The simplest sugars are glyceraldehyde and dihydroxyacetone. Depending on the total no. of carbon atoms present in the monosaccharide molecule, they are designated as below:

Number of Carbons	Name	Example
3	Trioses	Glyceraldehyde
4	Tetroses	Erythrose
5	Pentoses	Ribose
6	Hexoses	Glucose, Fructose
7	Heptoses	Sedoheptulose
9	Nonoses	Neuraminic acid

Individual Monosaccharides are discussed below:

1. Glucose = Dextrose = Blood Sugar

The parent monosacharide from which others are obtained and is also the basic building block of the most abundant polysaccharides-starch and cellulose. It is the most common carbohydrate. This is the main source of energy, most quickly absorbed by the human body. Glucose is the primary source of energy for brain. Glucose exists in the ring form, the body can metabolize only the D-isomer of glucose, the L isomer might be useful as an alternative sweetener. Relative sweetness of glucose is 0.7. This is found in ripe fruits, flowers, beetroot, honey etc.

2. Fructose = Levulose = Fruit Sugar

It is a structural isomer of glucose. Although it is a hexose, but can form either a five or six member ring. Natural sources of fructose include fruits, some vegetables, honey (half fructose and half glucose), sugar cane and sugar beets. This is absorbed much slower than glucose. After absorption by the small intestine and transport to the liver, fructose is almost all metabolized to glucose. However some fructose is converted to glycogen, lactic acid, or glucose depending on the metabolic state of the individual. Relative sweetness of fructose is 1.2-1.8.

3. Galactose

This is the last major monosaccharide of nutritional importance. It is structurally similar to glucose except the hydrogen and hydroxyl group on C-4 are reversed. This is not usually found free in nature in large quantities but rather combines with glucose to form a disaccharide called Lactose. Galactose in the diet is metabolized to glucose or glycogen. When later is required, as in the mammary gland of the lactating female, galactose is resynthesized using a wide variety of potential carbon atom sources.

4. Ribose

This is a five carbon sugar which is present in cell's genetic material and very little ribose is present in our diet. Human have the ability to produce ribose from the food we eat.

5. Monosaccharides Derivatives

☆ **Amino sugars-** sugars in which a hydroxyl group in the sugar is replaced by an amino group. Like glucosamine and galactosamine.

☆ **Deoxy sugars-** in hich one oxygen atom is removed at a designated carbon atom. Ex-L- Rhamnose, L-fucose.

☆ **Glycosides-** are the condensation products of sugar (glycose) with nonsugars (aglycoses). Some glycosides are cyanogenic.

☆ **Sugar Alcohols:** are obtained by hydrogenation of sugars using Raney nickel catalyst. Ex- Sorbitol, xylitol, and mannitol.

4.1.1.2. Oligosaccharides

These are formed by the polymerization of n molecules of monosachharides by the elimination of n-1 molecules of water. So, Oligosaccharides can be called as

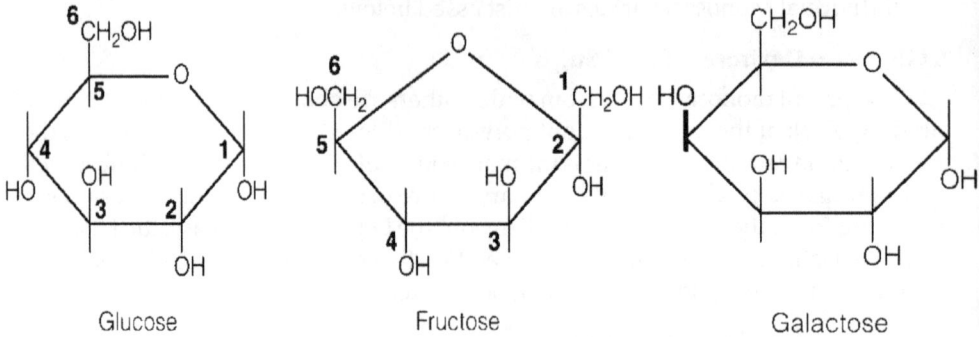

Figure 4.2: Structures of Monosaccharide.

hydrolyzable polymers of 2-10 monosaccharides. Oligosaccharide are linked by glycosidic linkages.

Disaccharides

These are formed when two monosaccharide combine by condensation reaction. In the Reaction, a water molecule is released and a glycosidic bond forms between the two monosacharides. Humans can digest such Carbohydrates only if the glucose molecules are linked by alpha glycosidic bonds. Examples of disaccharides are:

☆ Sucrose (glucose+fructose)

☆ Lactose (glucose+galactose)

☆ Maltose (glucose+glucose).

1. Maltose

Maltose is formed by condensation of two Glucose molecules joined by alpha bond. It doesn't occur in natural foods. It is found as an intermediate product by the action of enzymes, amylases on starch. It is of nutritional interest because it is a

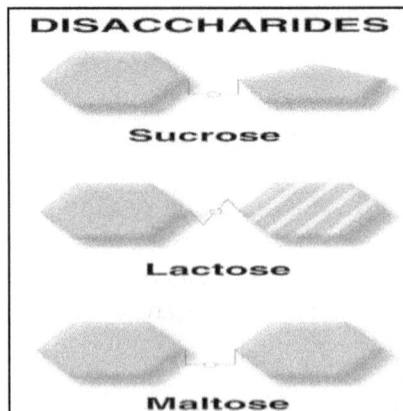

Figure 4.3: Structures of Disaccharides.

chemical intermediate in alcohol production in the beer and liquor industry. It is the sugar produced during the fermentation. Little maltose remains in the final fermentation product. Few products contain maltose. Digestion of starch in the mouth and small intestine yields maltose during the digestion process.

2. Sucrose

Sucrose is also known as common table sugar. It is composed of glucose and fructose molecule linked by alpha bond. Significant amount of sucrose are found only in plants such as sugar cane, sugar beets, honey and maple syrups. This can be purified to various degrees. For ex. Brown sugar, white sugar and powdered sugars are common forms of sucrose sold in grocery shop. In the presence of mild acids or the enzyme invertase, sucrose undergoes hydrolysis (Inversion). The hydrolysis product is known as **Invert Sugar**.

3. Lactose

This is the primary sugar of milk and milk products and is a constituent of mammalian milk. It consists of glucose joined to galactose with a beta bond. Many people are unable to digest large amounts of lactose, which can cause intestinal gas, bloating, cramping, and discomfort as the unabsorbed lactose is metabolized into acids and gases by bacteria in the large intestine. This is called as Lactose Intolerance.

Lactose Intolerance

Lactose intolerance refers to the inability to digest lactose results from a reduction in activity of the lactase enzyme. This enzyme is embedded within the surface of intestinal cells, splits lactose into glucose and galactose. These monosaccharides are absorbed from the small intestine into the bloodstreams, but lactose is not.

When lactase activity is low, lactose travel unaltered into the large intestine, where resident bacteria metabolizes it into acids and gases causing intestinal gas, bloating, cramping and discomfort. Persons suffering from lactose intolerance can eat hard cheese and regular yoghurt.

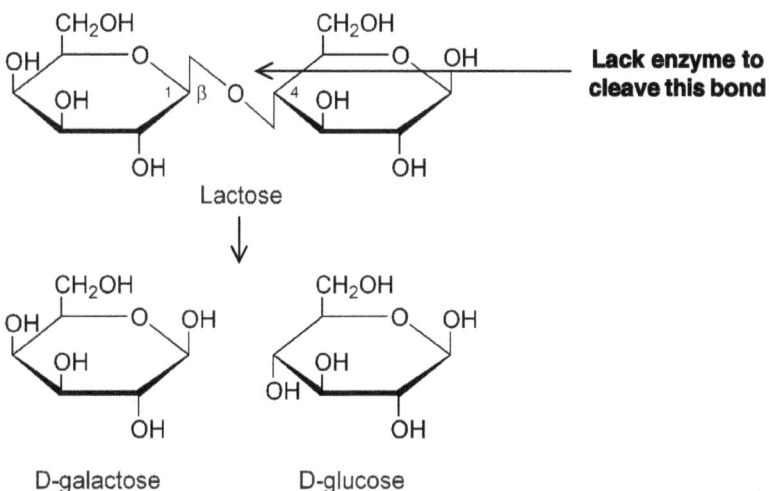

Lactose

D-galactose   D-glucose

4. Raffinose

This sugar is formed by the addition of one galactose residue to sucrose molecules. So it consists of 3 monosacchaides (galactose-glucose-fructose). This sugar is indigestible by the humans.

5. Stachyose

This sugar is formed by the addition of two galactose residue to sucrose molecules. So, it consist of 4 monosaccharides (galactose-galactose-glucose-fructose). This sugar is indigestible by the humans.

Raffinose and Stachyose

These sugars are found in beans and other legumes. These sugars remain undigested upon reaching the large intestine. There, bacteria metabolize them, producing gas and other by-products. Enzyme preparation (Beano) prevents the unpleasant side effects of intestinal gas if consumed right before a meal. Enzymes break down many of the indigestible oligosaccharides in legumes and other vegetables in the GI Tract before they reach the large intestine. This allows for the absorption of their monosaccharide components in the small intestine.

4.1.3 Functions of Glucose and other Sugars in the Body

1. Supply Energy to body cells.
2. Sparing protein from use as an energy source. If diet lacks enough carbohydrate to yield required glucose. The body is forced to make it from other nutrients such as proteins by the process called as gluconeogenesis.
3. These are involved in preventing ketosis. An adequate intake of carbohydrate (Glucose or other sugars or both) is necessary for complete metabolism of fats to CO_2 and water in the body. A low carbohydrate intake will result in decline in the release of insulin (hormone) leading to incomplete breakdown of fatty acids in the liver's metabolic pathways and to formation of ketones. Minimum carbohydrate level to prevent ketosis is 50-100g/day.
4. Improve palatability of foods.
5. Regulation of Blood Glucose.
6. Imparting flavour and sweetness to foods. Table 4.1 shows the relative sweetness of different sugars.

Table 4.1: Shows the Relative Sweetness of different Sugars

Sweetener	Relative Sweetness Sucrose=1	Typical Source
Lactose	0.2	Dairy products
Maltose	0.4	Sprouted seeds
Glucose	0.7	Corn syrup
Invert sugar	1.3	Some candies, honey
Sorbitol	0.6	Dietetic candies, sugarless gum
Xylitol	0.9	Sugarless gum
Aspartame	200	Diet soft drinks, diet fruit drinks, sugar less gum

4.1.2 Complex Carbohydrates/Polysaccharide

Polysaccharides are polymers (of > 10 monosaccharide's) containing many monosaccharide units upto 3000 or more. They are high molecular weight substances composed of a large number of monosacharide units combined to form one large molecule or polymer. They consists of a primary chains and in some side chains or branches may exist. Polysaccharide consists of primary chain of 1-2 types of monosaccharide units and the side chain consists of sugars different from those of the main chain with different types of glycosidic linkages.

They provide structural material (cell walls, fibres, seed coats, peels and husks) for example cellulose, hemicellulose, pentosans and pectic substances. Chitin and mucopolysaccharides serve this purpose in animals. Structural polysaccharides are indigestible substances even then they are important for human health. They provide the bulk in the diet and aid excretion.

Starch, dextrins and fructan in plants, glycogen in animals provide food reserves. These polysaccharides are digested and utilized by the human body. They are nutritionally important. Polysaccharides attract and retain water so that life's enzymic processes are not impeded under dehydrating conditions. The generic name of polysaccharide is GLYCAN. Polysaccharides are classified on the basis of structure and digestibility as below:

4.1.2.1 Based on Structure

1. **Homopolysaccharides:** are polymers of a single type of monosaccharide, for ex: Glycogen, Cellulose.
2. **Heteropolysaccharides:** are polymers of at least 2 types of monosaccharide, for ex: Glucosaminoglycans.

4.1.2.2 Based on Digestibility

1. **Digestible** – Starch and Glycogen
2. **Indigestible-** Cellulose-Dietary Fiber

4.1.2.2.1 Digestible Polysaccharides

1. Starch

Starch is a digestible plant polysaccharide, which contain only glucose residues. This is the principle food reserve polysaccharide, it occur in cereal grains, pulses, tubers, bulbs and fruits in varying amount. It provides major source of energy in the diet of human.

They are mixtures of structurally distinct polysacharides- There are two types of plant starch- amylose (25 per cent) and amylopectin (75 per cent). Waxy or glutinuous starches like waxy corn starch contain little or no amylose.

a. Amylose

It consists of glucose units linked by alpha (1-4) bonds, chain length varies and it is considered to have a molecular weight of 1.1 to 1.9 million. Amylose is not truly soluble in water but forms hydrated micelles. It is this structure of amylose that is responsible for the blue colour produced by iodine with starch. It contribute to the gelling characteristics of cooked and cooled starches. It has great industrial possibilities.

b. Amylopectin

It has a backbone of alpha (1-4) linkage but in addition, the molecule is branched through alpha (1-6) bonds to the extent of 4-5 per cent. The length of the linear units in amylopectin is about 20-25 glucose units then branched unit. Molecular weight of amylopectin is over 10 million. It has a brush like structure. It is responsible for thickened properties of starch preparation but doesn't contribute gel properties.

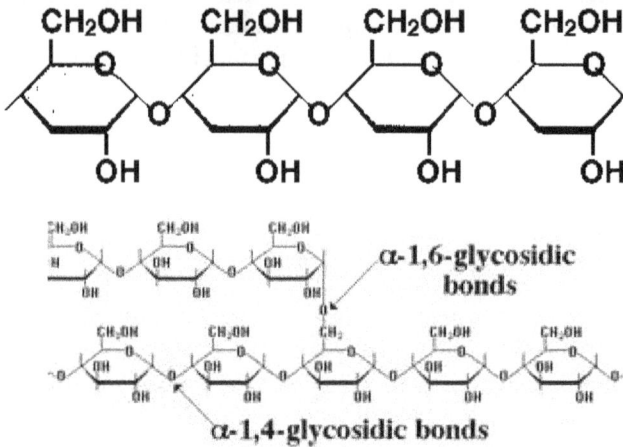

Figure 4.4: Structure of Starch (Amylase and amylopectin respectively)

2. Glycogen (Sometimes called as animal starch)

It is known as storage polysaccharide in animals. This is mainly present in liver and skeletal muscle. It is a branched chain polysaccharides resembling amylopectin rather than amylose but is more branched than amylopectin. Branching is at every 4-8 glc residues (alpha 1-6). Liver glycogen rapidly hydrolyzes to D-glucose after slaughter of animals. Liver and Muscle are the major storage sites for glycogen. Highly branched and composed of multiple glucose molecules.

Starch from plants is hydrolysed in the body to produce glucose. Glucose passes into the cell and is used in metabolism. Inside the cell, glucose can be polymerised to make glycogen which acts as a carbohydrate energy store.

Enzymatic breakdown of glycogen yields glucose phosphate molecules. The liver contains the enzyme that converts this glucose phosphate to glucose, which can then enter the bloodstream, but muscles lack this enzyme. Thus liver glycogen plays an important role in regulation of blood glucose. Although muscle glycogen isn't converted to blood glucose it does supply glucose for muscle use especially during high intensity and endurance physical activity.

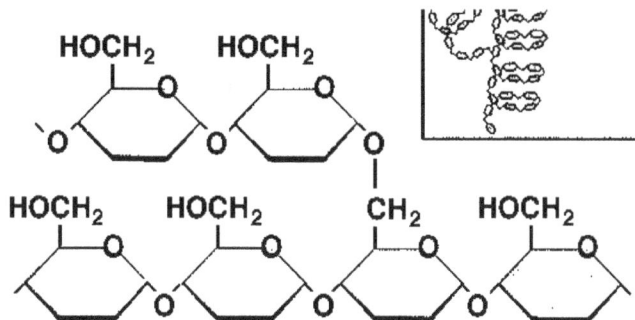

Figure 4.5: Structure of Glycogen.

3. Dextrin

These are products of partial breakdown of starches. These are the intermediate in size between starches and sugars and exhibit properties that are intermediated between these classes of materials. These are formed when starch is subjected to dry heat. For ex. Tosting of bread converts a part of starch to dextrin. The sweet taste of toast is due to dextrinization.

4. Cellulose

Cellulose is a polymer made from glucose. But it's made from β-glucose molecules and the polymer molecules are 'straight'. Cellulose makes up more than 25 per cent of cell all in higher plants. Cellulose molecules are associated into partially crystalline microfibrils. Cellulose is found embedded in an amorphous gel composed of hemicellulose and pectic substances. Cellulose has very high molecular weight. It doesn't dissolve in water to any extent. It is not digested in the human system and acts as roughage or dietary fibre. It can be broken down to glucose by certain microbial enzyme. (This is the way cellulose is hydrolyzed in the rumen of animals to provide energy).

Changes in cell wall constituents surrounding cellulose and in the interstitial matrix is largely responsible for the changes in texture of fruits and vegetables during ripening, storage and processing.

Thus, cellulose is not digested by humans and is classified as a dietary fiber, not a starch. The long glucose chains of cellulose are linear and they can pack very

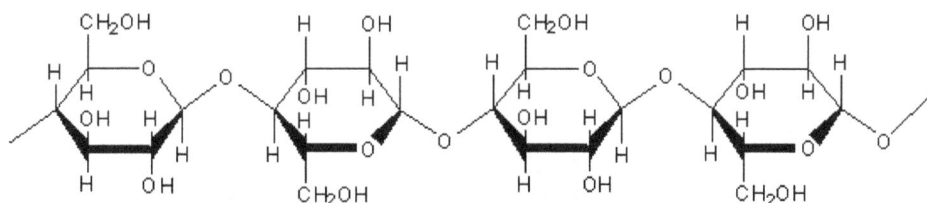

Figure 4.6: Stucture of Cellulose.

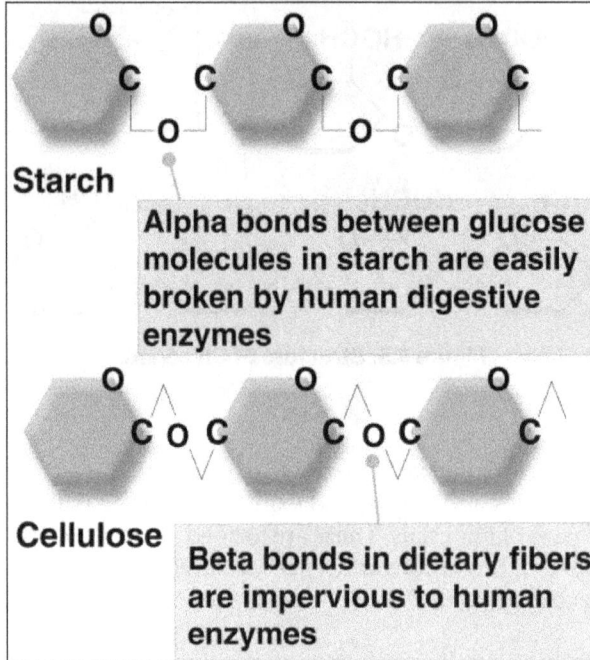

Figure 4.7: Showing the difference in the Structure of Starch and Cellulose.

closely forming fibrous structure of great strength. Examples and major sources are mentioned below:

Pure cellulose is found in cotton ball,

Bran fiber is rich in hemicellulose,

Whole grains are rich in dietary fiber,

Woody fibers in broccoli, cabbage are rich in lignins.

5. Pectic Substances

They are mixture of polysacchaides formed from galactose, arabinose and galacturonic acid. They occur as constituents of plant cell walls and in the middle lamella.

6. Gums

Most gums are polysaccharides. They function as thickeners for gravies and sauces, moisture retention agents in baked goods. Examples include Seed gums, plant exudates and seaweed gums.

4.1.2.2.2 Indigestible Polysaccharide–Dietary Fiber

These refer to substances in food (essentially from plants) that cannot be broken down by the normal digestive processes in the stomach and small intestine. These substances add bulk to the feces. Some are metabolized by bacteria in the large intestine. These are classified as soluble and insoluble dietary fiber as discussed below:

Soluble Dietary Fiber	Insoluble Dietary Fiber
Fibers that either dissolve or swell in water and are metabolized (fermented) by bacteria in the large intestine.	Fibers that mostly do not dissolve in water and are not metabolized (Non fermentable) by bacteria in the large intestine.
These are partly fermentable by bacteria in the large intestine into the forms which can be used.	Non fermentable by bacteria in the large intestine.
These include pectins, gums, and mucilages. Rich sources of soluble fibers are fruits and vegetables, soybean fiber, rice bran. **Psyllium** is mostly soluble type of dietary fiber found in the seeds of the plantain plant.	These include cellulose, some hemicellulose and lignins.
These finds uses in jams, jellies, ice creams, salad dressing etc.	
Health effects: These fibers lower blood cholesterol, slower down glucose absorption and transit of food through upper digestive tract. They hold the moisture in stool and soften it.	**Health effects:** These fibers soften stoold, regulate bowel movement and speeds up the transit of material through the small intestine thus increasing the fecal weight and speed of fecal passage through colon, thus reducing the risk of diverticulosis, hemorrhoids, appendicitis and even reduce the risk of colon cancer.

How Dietary Fiber are Metabolized in the Body?

1. Bacteria in the large intestine metabolize soluble dietary fibers into products such as short chain fatty acids (acetic acid, butyric acid, propionic acid) and gases.
2. These acids (esp. Butyric acid) provide fuel for the cells in the large intestine and enhance their health.
3. These products can also be absorbed into the bloodstream. As a result of bacterial metabolism, soluble dietary fibers yield about 3kcal/g on average.
4. When intake of dietary fiber is high, its breakdown by bacteria can cause certain gases such as methane and hydrogen.

What is Crude Fiber?

The term arose during early 1900's to reflect the amount of indigestible foodstuff present in animal feed. Crude fiber consists primarily of cellulose and lignin. Crude fiber can be quantified by boiling the sample for 1 hour in acid and for another hour in an alkaline solution. The remains of that chemical digestion was called Crude Fiber.

4.2.3 Health Benefits of Dietary Fiber

1. Supplies mass to the feces, making elimination much easier as plant fibers attract water, stool will be large and soft. Large size stimulates the intestinal muscles which aids elimination. Consequently, less pressure is necessary to expel the stool.

Opposite can happen when too little dietary fibers are consumed, the stool may be hard and small. Constipation can result which can force the individual to exert excessive pressure in the large intestine during defecation. And this high pressure can force parts of the large intestine wall out from between the surrounding bands of muscles forming small pouches called Diverticula. 80 per cent of diverticulitis is asymptomatic in affected people. The asymptomatic form of this disease is called diverticulosis. If the diverticula become filled with food particles such as seeds, hulls, bacteria can metabolize these food particles into acids and gases. The acids and gases irritate the diverticula and may eventually cause them to become inflamed, and the condition is known as diverticulitis. After inflammation subsides, High fiber diet can be given for easy stool elimination.

2. When consumed in large amounts, soluble dietary fibers can slows glucose absorption from the small intestine. It is helpful in treatment of diabetes.

3. A high intake of soluble dietary fibers also inhibits cholesterol absorption from the small intestine, thereby reducing serum cholesterol.

How Does Soluble Fiber Reduce Cholesterol?

There are two ways by which soluble fiber reduce cholesterol:

   a. Enterohepatic circulation of Bile.
   b. Bacterial by-products of fiber fermentation in the large intestine inhibit cholesterol synthesis in the liver.

4. High fiber diet will lower glucose absorption thus will reduce release of insulin, this reduction in insulin may contribute to the ability of soluble fibers to lower serum cholesterol.

5. Helpful in weight control and reduces the risk of developing obesity. This happens because the bulky nature of high fiber foods fills us up without yielding much energy.

6. Prevent cancer of large intestine (Colon cancer).

7. Hemorrhoids may also develop due to excessive straining during defecation.

8. Vitamin C and Carotenoids from fruits and vegetables and reduction in meat and fat intake may exert the main protective effect than dietary fiber alone.

9. A low calcium intake is also implicated in colon cancer risk.

What are the Problems with High Fiber Diet?

1. Develop Phytobezoars (Pellets of dietary fiber in stomach). Mainly found in diabetic and elderly people who consume large amounts of dietary fibers.

2. May binds with Iron and Zinc and make them unavailable, if consumed in excess.

4.2 Recommended Carbohydrate Intakes

☆ As such No RDA has been established for carbohydrate. RDA is given for calorie =2200kcal (source USDA). RDA values are given for carbohydrate on the basis of energy we get from carbohydrate.

❖ Infant = 40-50 per cent of calorie from carbohydrate,

❖ Adults and adolescences=50-70 per cent,

❖ Nursing Mother or expecting = 40-60 per cent calorie.

❖ Overall the requirement can be high but the fulfillment of the requirement is not mandatory by carbohydrate it can be fulfilled by other sources.

☆ But, it is important to consume atleast 50-100g of carbohydrate/day to prevent ketosis. This is the minimum carbohydrate intake continues to spare protein and helps meet energy needs.

☆ Nutritional experts recommend that we should include more starch and fiber in our diets. Goal should be set to get 45 per cent of energy intake, with total carbohydrate intake accounting for about 55-60 per cent of total energy intake.

For Dietary fiber, a reasonable intake is 20-35g/day (10-13g/1000kcal). Fiber intake of 20-35g should prevent diverticulosis. Eating a high fiber cereal for breakfast is a good alternative by including whole grains, fruits, vegetables in the diet.

☆ Moderate intake of simple sugars should provide no more than 10 - 15 per cent of total energy intake daily. This corresponds to maximum value of 75 per cent.

☆ Lowe Income Group people derives much of the energy requirement from carbohydrate.

4.3 Functions of Carbohydrate

1. They provide energy and serve as energy resource. They are the sources of metabolic fuels and energy reserve.

2. These are structural components of cell wall. In plants and of the exoskeleton of antropods (insect family).

3. Carbohydrates can also be used in making of nonessential amino acids, fats, components of DNA, RNA (deoxy and ribose sugar) and other structures and compounds in the body.

4. These are integral features of many proteins and lipids (glycolipids and glycoproteins). They are present in cell membrane.

5. Provide sweetness and humectant properties to food (to keep the product moist).

6. Galactose is component of nerve membrane.

7. Glucose is needed by brain 120g/day.

8. Add bulk to the diet (dietary fibers).

4.4 Sources of Carbohydrate

1. Cereals and millets,
2. Legumes and Pulses,
3. Fruits and Vegetables.

4.5 Carbohydrate Digestions and Absorption

Glucose is the basic carbohydrate unit that helps fuel all the cells of the body. During digestion the body must be equipped to convert any mono-, di- or polysaccharide into usable form of glucose.

1. The Mouth and Stomach

☆ Digestion begins in the mouth,

☆ In the mouth glands (parotid, submandibular, and sublingual glands) produce saliva.

☆ Chewing action in the mouth mechanically breaks food down in to smaller particles, allowing more surface area for efficient digestive action by enzymes (amylases).

☆ Saliva also contain lysozyme (enzyme that kill bacteria by rupturing cell membranes)

☆ Amylase is responsible for breaking the long chains of glucose molecules in starches into smaller units. The resulting products are maltose and smaller starch chains known as dextrins.

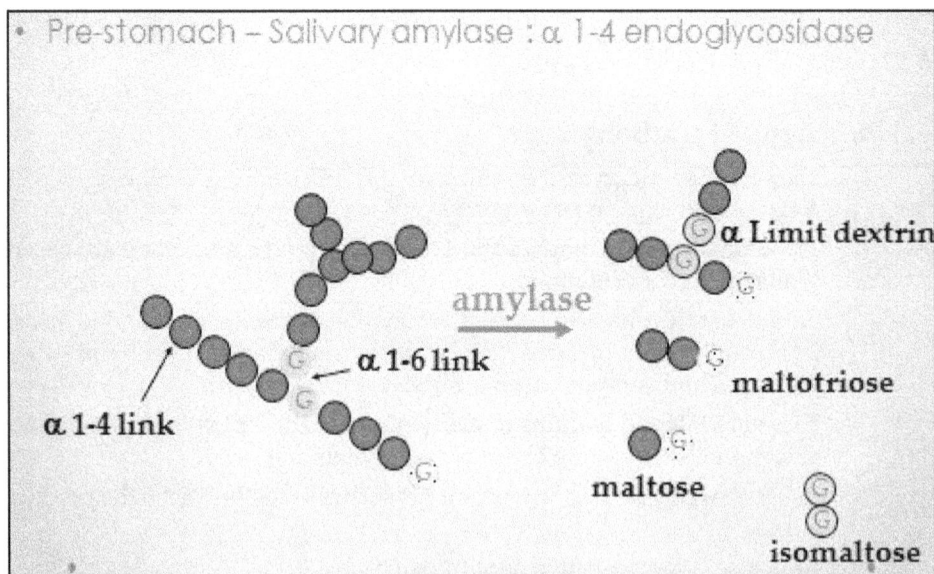

Figure 4.8: Digestion.

✫ The food doesn't remain in the mouth for long before it is swallowed and moves into the stomach.

✫ The food exits the esophagus and enters the stomach (holding tank of about 4cups (1L). Food remains here for 2-3 hours. High fat meals take the longest time to empty. The high acidity of the stomach deactivates amylase, so no further digestion of carbohydrate takes place until food moves into the small intestine.

Stomach

✫ Not much carbohydrate digestion

✫ Acid and pepsin to unfold proteins

✫ Ruminants have forestomachs with extensive microbial populations to breakdown and anaerobically ferment feed

2. Small Intestine

✫ Small intestine has three sections duodenum (10 inch long), jejunum (4 feet long), ileum (5 feet). Food remains for about 3-10 hours. Small intestine is considered as small because of narrow diameter (1inch).

✫ Most digestion is completed in the duodenum and upper jejunum with the help of enzyme made by intestinal cells and pancreas (pancreatic amylases).

✫ As carbohydrate enters the small intestine, pancreases releases its amylases to continue the digestion of starches through the dextrin phase into maltose. For examples, when a fruit shake is taken in the body, the carbohydrate of

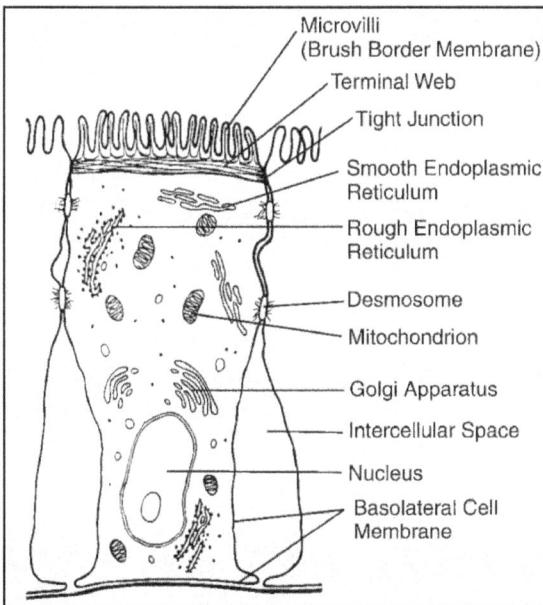

Figure 4.9: Showing the Ultra Structure of Small Intestine Epithelial Cell (Enterocyte).

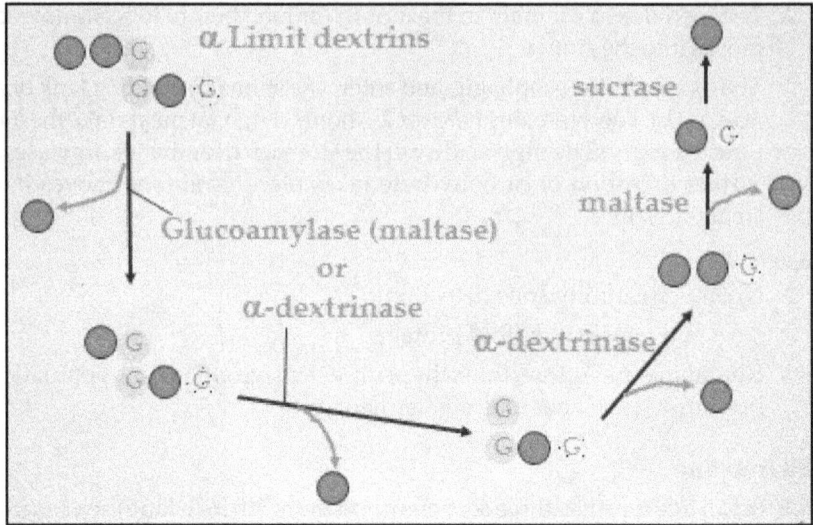

Figure 4.10: Oligosaccharide Digestion.

fruit shake are now present as monosaccharide (glucose, fructose) and disaccharides (maltose, lactose, sucrose).

☆ Eventually all disaccharides are digested to monosaccharide's by disaccharidases (these are synthesized by and attached to, the cells of the small intestine).

☆ Maltose is acted on by maltase to produce- 2 glucose molecules. Sucrose acted upon by sucrases to produce glucose and fructose. Lactose by lactase to produce glucose and galactose. However without lactase, lactose remains in the digestive tract and is fermented by microbes in the lower intestine producing gas, causing abdominal bloating, flatulence and cramps. The single sugars (like glucose) are the form of carbohydrates that can be absorbed.

3. Large Intestine

☆ Not all carbohydrate is absorbed into the body. Fiber, which is indigestible, proceeds through the digestive tract into the large intestine, or colon.

☆ Some of this fiber is broken down by bacteria and absorbed.

☆ Cellulose, lignins and some hemicelluloses have properties of holding onto water and swelling, resulting in a stool that feels large and stimulates peristalsis, the wave like muscular action of the digestive tract which is responsible for moving the contents along in the large intestine.

☆ The water insoluble fibers, therefore, help to move the stool through faster.

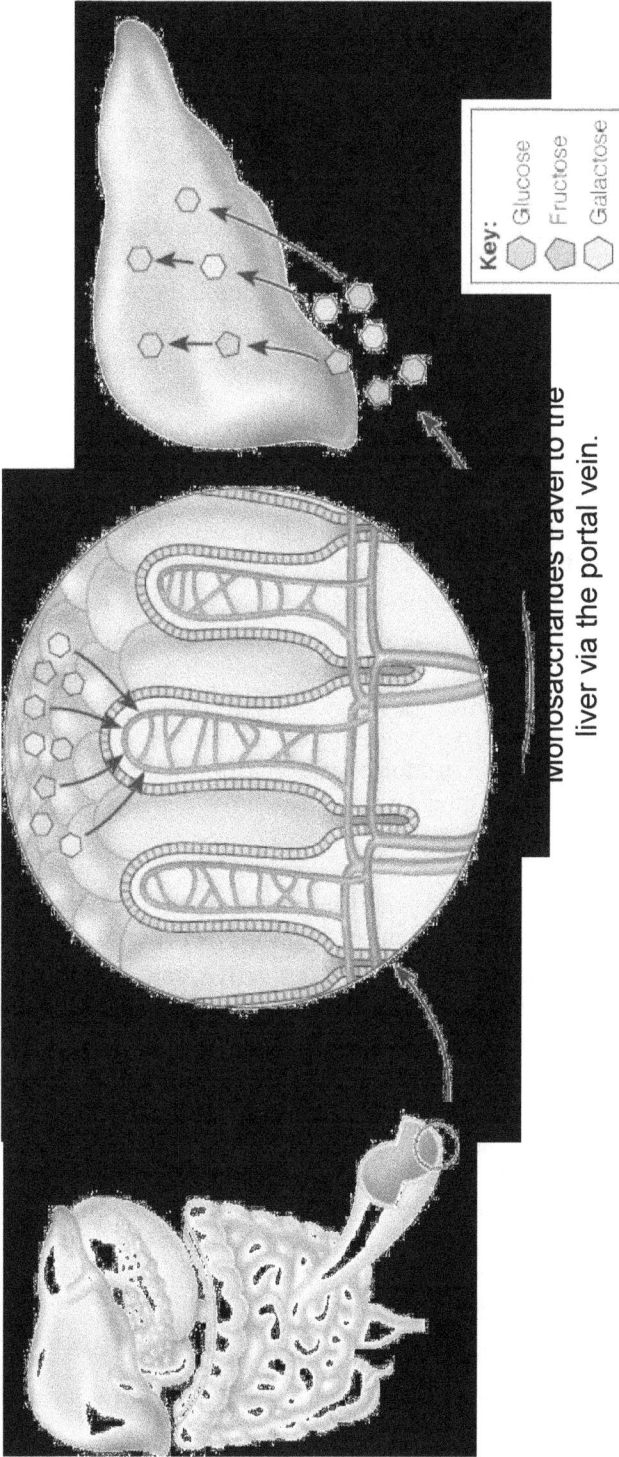

Monosaccharides, the end products of carbohydrate digestion, enter the capillaries of the intestinal villi.

Monosaccharides travel to the liver via the portal vein.

Key:
Glucose
Fructose
Galactose

Figure 4.11

Carbohydrate Absorption

☆ Small intestine is the site where all nutrients are absorbed. Absorption of carbohydrate takes place as monosaccharide move across the intestinal lining and into the bloodstream.

☆ Absorption can be slowed by the presence of water soluble fibers, which makes it difficult for enzymes to act quickly in breaking down polysaccharides.

☆ Disaccharides are hydrolyzed by specific diasacharidases like sucrases (Sucrose), Maltase(maltose) and lactase (lactose), which are released into the lumen from the epithelial cells, lining the walls of small intestine.

☆ The major absorptive sites are fingerlike projections in the small intestine called villi. The absorptive cells that cover the villi have a life span of 2 to 5 days. Thus the intestinal lining continually renew itself. Absorptive cells can perform passive, facilitated and active absorption, as well as endocytosis.

☆ Fructose is taken up by the absorptive cells via facilitative absorption. Glucose and galactose are actively pumped into the absorptive cells of the villi along with sodium during active absorption.

☆ Once glucose, galactose and fructose enters inside the villi, they are transported via the portal vein to the liver for glycogen production, fat production, energy use, or direct release into the blood.

☆ Presence of Phytic acid, lectins and polyphenols (tannins) (plant chemicals (phytochemicals) acts as amylase inhibitors) especially in legumes. They slow down the digestion of carbohydrate in the digestive tract resulting in flattened blood glucose response curves.

Conclusion

☆ A mixture of simple sugars mainly hexoses is absorbed into the epithelial cells of the intestinal walls and brought via portal blood to liver.

☆ Simple process of diffusion and active transport of sugars (movement against concentration gradient, in energy derived process) are the mechanisms by high sugars are absorbed.

☆ Monosaccharides are still too large for passive diffusion across brush border membrane. Body use facilitated diffusion to absorb these molecules. Glucose and galactose use a sodium-glucose symport (SGLUT1) while fructose uses the glut5. Sodium must be transported out of the cell to maintain proper electrochemical gradient (sodium potassium pump). Water will also follow sodium into enterocyte. This is critical to maintain proper water balance.

☆ Two main mechanisms of absorption are observed as mentioned below:

1. Facilitative transport: with the gradient.

2. Active transport : against the gradient.

1. **Facilitative transport:** Glucose and other hexoses could be transported by facilitative transporters.

2. Active transport: Glucose is actively transported against gradient, coupled with Sodium; the former against the gradient and the latter with the gradient.

Absorptive Disorders of Carbohydates

1. **Lactose intolerance** may be attributable to lactase deficiency, which is a rate limiting enzyme for Lactose absorption.It is different from milk intolerance which is sensitivity condition to milk protein, usually β-lactoglobulin.

2. **Sucrase deficiency:** Under this condition both sucrase and isomaltase are deficient as they coexist as a complex enzyme. The symptoms appear from early childhood.

3. **Disacchariduria:** This is due to disaccharidase deficiency, more than 300 mg of disaccharides appear in urine also occur in patients with intestinal damage.

4. **Monosaccharide malabsorption:** Symptoms like diarrhea, flatulence, abdominal cramps occur in patients.

4.6 Concern with Eating Carbohydrate Rich Food

4.6.1 Concern with Eating Simple Sugars

a. **Dental caries**: are formed when sugars and other carbohydrates are metabolized into acids by bacteria that live in the mouth. *Streptococcus mutans* dissolves the tooth enamel and underlying structure. The long lived carbohydrates are termed as cariogenic. Liquid sugar sources (fruit juices) are not potent source causing dental caries as sticky or gummy foods.

4.6.2 Concern with eating starch containing foods

b. If starch containing foods are held in the mouth for long time then they can be acted on by enzymes in the mouth that break down the starch to sugars and further to acids by bacteria.

c. Type II Diebetes Mellitus.

4.6.3 Carbohydrate intake related disorders

1. **Lactose intolerance**- Diarrhoea and generation of gases due to fermentation of undigested lactose by intestinal microorganisms.

2. **High blood sugar – diabetes**, discussed in section 4.7

3. Obesity

4. Cancer: A diet low in dietary fiber

5. Cardiovascular Disease

6. Gastrointestinal Disorders

7. With high fiber diet

   7.1. Develop Phytobezoars

   7.2. May binds with Iron and Zinc and make them unavailable.

Summary of Carbohydrate Digestion and Absorption

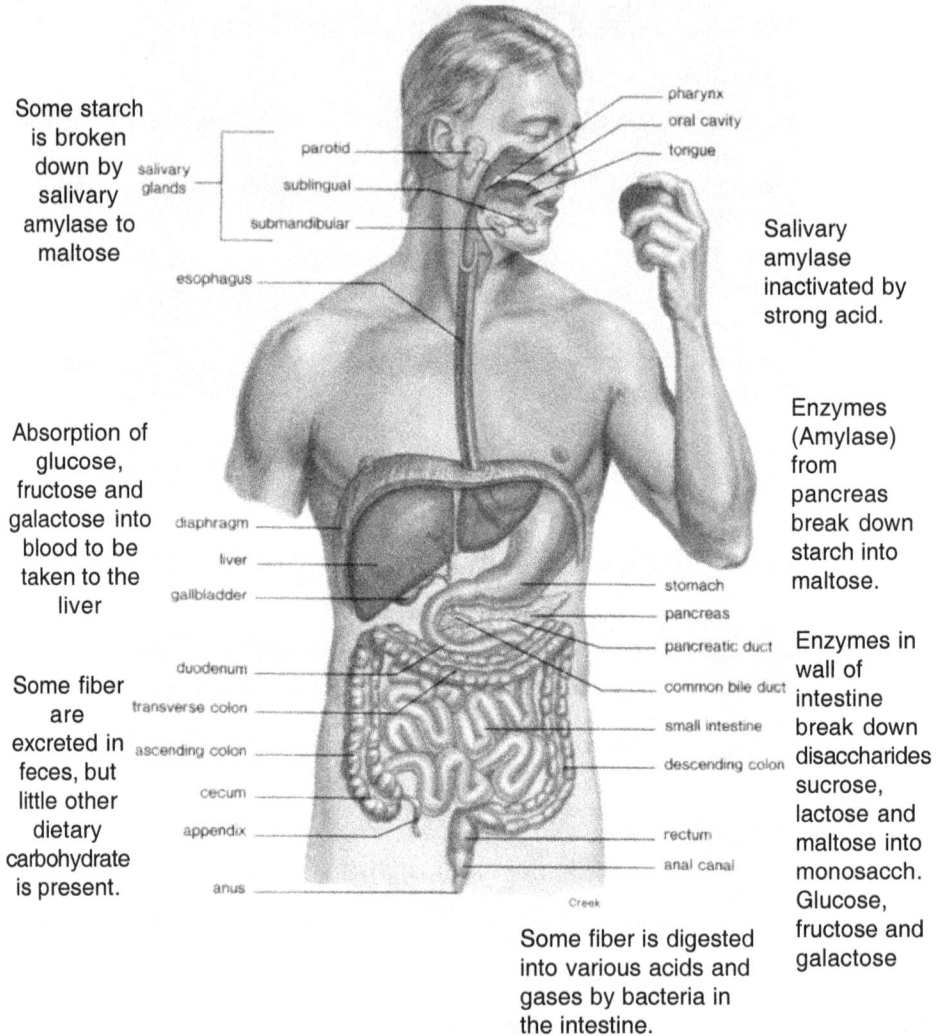

Some starch is broken down by salivary amylase to maltose

Salivary amylase inactivated by strong acid.

Absorption of glucose, fructose and galactose into blood to be taken to the liver

Enzymes (Amylase) from pancreas break down starch into maltose.

Some fiber are excreted in feces, but little other dietary carbohydrate is present.

Enzymes in wall of intestine break down disaccharides sucrose, lactose and maltose into monosacch. Glucose, fructose and galactose

Some fiber is digested into various acids and gases by bacteria in the intestine.

pharynx, oral cavity, tongue, parotid, salivary glands, sublingual, submandibular, esophagus, diaphragm, liver, gallbladder, stomach, pancreas, pancreatic duct, common bile duct, small intestine, descending colon, duodenum, transverse colon, ascending colon, cecum, appendix, rectum, anal canal, anus

Figure 4.12: Showing the Digestion and Absorption Process of Carbohydrate.

8. Dental caries.
9. Marasmus- Energy as well as protein deficiency.
10. Ketosis and gluconeogensis will occur if sufficient amount of carbohydrates is not supplied in diet.
11. Some researchers have suggested that sucrose affects behavior, especially in children. They claim sucrose creates an excited, even antisocial state, which may lead to violence and disruptive behaviour.

4.7 Regulating Blood Glucose Levels

There are mainly three hormones which regulate the blood glucose level within the n**ormal Blood Glucose range** within 70-115mg/100ml of blood in the fasting state. Hormones regulating the blood glucose level are namely **Insulin –** which lowers blood glucose level, **Glucagon –** which raises blood glucose and **Epinephrine –** which is a "fight-or-flight" hormone.

☆ **Hyperglycemia** refers to a condition of high blood glucose, above 140mg/100ml of blood.

☆ **Hypoglycemia** refers to a condition of Low blood glucose, below 40- 50mg/100ml of blood. **Fasting Hypoglycemia** refers to low blood glucose that follows about a day of fasting.

☆ **Reactive Hypoglycemia (Postprandial hypoglycemia)** refers to a condition of low blood glucose that follows a meal high in simple sugars, with corresponding to symptoms of irritability, headache, nervousness, sweating and confusion.

High Blood Sugar – Diabetes

Diabetes is a disorder in which enough insulin is not present in the bloodstream to move glucose into the cell and use it for energy. As a result of the ineffective levels of insulin, blood glucose rises. The symptoms and complications of diabetes are caused by this high blood glucose. There are two major types of diabetes mellitus: Insulin dependent diabetes and non-insulin dependent diabetes.

☆ Type I – Insulin Dependent Diabetes Mellitus (IDDM) or Juvenile Onset

☆ Type II – Non Insulin Dependent Diabetes Mellitus (NIDDM) or Adult Onset

1. Type –I or Insulin Dependent Diabetes Mellitus (IDDM)

In IDDM, the pancreas of individual doesn't make insulin. Without insulin, the body cannot burn glucose for energy and try to use body's reserve of fat alone. This results in buildup of ketone in the blood supply, resulting in ketoacidosis. If this condition remains untreated then this can even results in death. This type of diabetes usually occurs in young children or adolescents and is also known as Type I or Juvenile onset diabetes. In this case individuals require an external source of insulin to sustain life. Only around 10 per cent of all diabetes are diagnosed with IDDM. Risk factors of type-I is more linked with Genetic disorder.

2. Type –II or Non Insulin Dependent Diabetes Mellitus (NIDDM) or Adult Onset

The pancreas of individuals with Non Insulin Dependent Diabetes Mellitus (NIDDM) make some insulin, but it is either insufficient or is not effective. While these individuals are not prone to the life threatening ketoacidosis, they may still require insulin form an external source for correction of high blood glucose if this cannot be achieved by diet and exercise. Of all the cases of diabetes, 85 per cent are NIFFM, it usually occurs after the age of 40 and 85 per cent of those diagnosed are obese.

Though obesity (excessive accumulation of body fat) increases the risk of diabetes, it is also seen in adults who are not obese, but who may have central (abdominal) obesity. In other words, centrally distributed abdominal body fat (abdominal adiposity) is also known to increase insulin resistance and diabetes type –II. In subjects with this type of diabetes, diet control, physical activity and oral anti-diabetic drugs may be sufficient to control the increased blood glucose.

Prevalence

It is estimated that about 9 per cent Indians suffer from Diabetes. Prevalence of diabetes is 2-3 fold higher in urban than in rural India and occurs at a much younger age in Indians. More than 90 per cent of all diabetes belongs to type -II diabetes. It is generally believed that diabetes is more prevalent in affluent societies where obesity is a major health problem. However, diabetes exists in all populations with variations in prevalence between ethnic groups and geographic areas. Despite the wide prevalent under nutrition, India currently has 62.4 million diabetics (ICMR study). Based on current rates, prevalence of diabetes in India is expected to reach 100 million by 2030.

Type 2 Diabetes is Preventable

Physical activity and diet regulations are important factors to control diabetes. University of Michigan Health System show that men and women who walked for 30 minutes five days a week, decreased their fat and total calorie intake, and reduced their body weight by 7 per cent over a period of three years were able to cut their risk of developing type 2 diabetes by 58 per cent.

Symptoms

Many diabetics may not be aware that they have diabetes. The following are the symptoms in an otherwise healthy person should make a person to go to doctor for knowing if he or she may have diabetes:

- ☆ Loss of weight
- ☆ Increased appetite and thirst
- ☆ Frequent urination (polyuria)
- ☆ Lethargy and Sleepiness
- ☆ Slow healing of cuts and wounds

Risk Factors for Type 2 Diabetes

Why do we develop type II diabetes?

1. **Body Mass Index (BMI):** BMI is an excellent indicator of the weight status of any person. It is defined as the weight (in kg) divided by height (in m^2). A healthy BMI for Indian adults is between 19 and 22. Indians are unique as we are at risk of developing type II diabetes at a lower BMI. We, in India may develop type II diabetes at much younger age and at lower BMI. Diabetes is a common consequence of overweight and obesity in adults. Obesity is also a strong risk factor.

2. **Waist to Hip Ratio (WHR):** Shape of the body more than the body's weight is an important factor in determining risks for the development of many

diseases such as diabetes, high blood pressure, lipid disorders, and atherosclerosis leading to cardiovascular diseases (CVD) and stroke. WHR can be measured by:

1. Measure waist at navel while standing relaxed, not pulling in your stomach.
2. Measure around hips, over the buttocks at the widest part of the buttocks.
3. Divide the waist measure by the hip measure; this gives waist hip ratio (WHR).

Research shows that people with "apple-shaped" bodies (with more weight around the waist) are at higher risk than those with "pear-shaped" bodies that is more weight around the hips. Waist-Hip Ratio (WHR) is used along with BMI for predicting risk for many of the serious disorders such as diabetes, high blood pressure etc. Waist to Hip Ratio of more than 0.80 is high risk for women and more than 0.95 is high risk for men.

3. Physical activity: Physical activity plays an important role in the treatment of diabetes. Regular exercise can increase the sensitivity of the body tissues to insulin by more than 30 per cent. Many studies suggest that walking briskly for a half hour every day reduces the risk of developing type 2 diabetes by 30 per cent. Physical activity helps to correct the associated lipid abnormality, increases cardiovascular fitness as well. However long hours of sweating exercise are not required to keep blood sugar under control.

4. Healthy eating habit can prevent diabetes: Our eating habits can affect our blood sugar levels. Foods containing refined flour, whether as starches or sugars, will immediately raise blood sugar and increase insulin resistance. So, a diabetic should avoid maida rich products like pastries, biscuits and cakes. Also, they should consider eating brown rice or whole wheat flour product over polished white rice or refined flour/maida respectively.

A diet high in fat or calorie, or high in omega-6 fatty acids (n–6) and low in omega-3 fatty acid (n–3) will cause inflammation, this further can cause metabolic syndrome, insulin resistance, type 2 diabetes, coronary heart disease, metabolic dysfunction, prothrombotic processes, stroke and alzheimer's disease etc. Whole and fresh fruits and green leafy vegetables should be preferred over fruit or vegetable juices to reduce inflammation because they are low in starchy carbohydrates and full of fiber, minerals and vitamins. Correct ratio of omega-3 and omega-6 fatty acids is important to main good health. Nuts (Walnuts, almonds, hazelnuts etc.) are rich in minerals, fibers, vitamins and omega-3 fatty acid. Eating nuts as part of a healthy diet is good. The calorie coming from nuts should be judiciously compensated by decreasing calorie intake from cereals or fats.

5. Yoga or meditation will help overcome stressful situations: Stressful situations can release hormones that will raise blood sugar.

6. Smoking and alcohol: Smoking and alcohol increase risk for insulin resistance, which often leads to diabetes. In diabetics, it increases the risk of complications, which include heart disease, stroke and circulation problems.

Diagnostic Criteria for IFG, IGT and Diabetes

Different Types	Plasma Glucose Level (mg/dl)	
	Fasting	2-hr post-load
Normal	<110	<140
Impaired fasting glucose (IFG)	110 to125	<140
Impaired glucose tolerance (IGT)	<126	≥140 and 200
Diabetes	≥126	≥200

Consequences of Diabetes

☆ Diabetes give rise to peripheral vascular disease in which the extremities of the body don't get enough blood, leading to the tissue death.

☆ Deterioration of the eye. Around 80 per cent of diabetic individuals have some degree of retinopathy. This is a hemorrhage in the capillaries of the retina of the eye.

☆ Diabetes is also responsible for kidney, heart disease and progressive nerve damage.

☆ Diabetic women who are pregnant have a higher risk of infant mortality, birth defects, respiratory problems, prematurity and other health disorders.

☆ Individuals with diabetes suffer from coronary artery disease, and are at twice to thrice at the risk of suffering heart attack or stroke than those without disease.

What Type of Nutrition should be Followed in Diabetes?

There are four nutritional goals of managing diabetes.

1. Maintain appropriate blood glucose levels.
2. Achieve and maintain reasonable weight.
3. Achieve and maintain healthy blood lipid levels.
4. Practice good nutritional habits.

The nutrition goals can be achieved by eating less fat and more complex carbohydrate. Figure 4.13 lays the summary of diabetes. Diabetic person should avoid sugar but in case they want to have sugar in tea or any other food then they can take alternative natural sweetener. Sugar Replacers can also be opted as they give less energy than sugar as mentioned below:

☆ Isomalt: 2.0 kcal/g
☆ Lactitol: 2.0 kcal/g

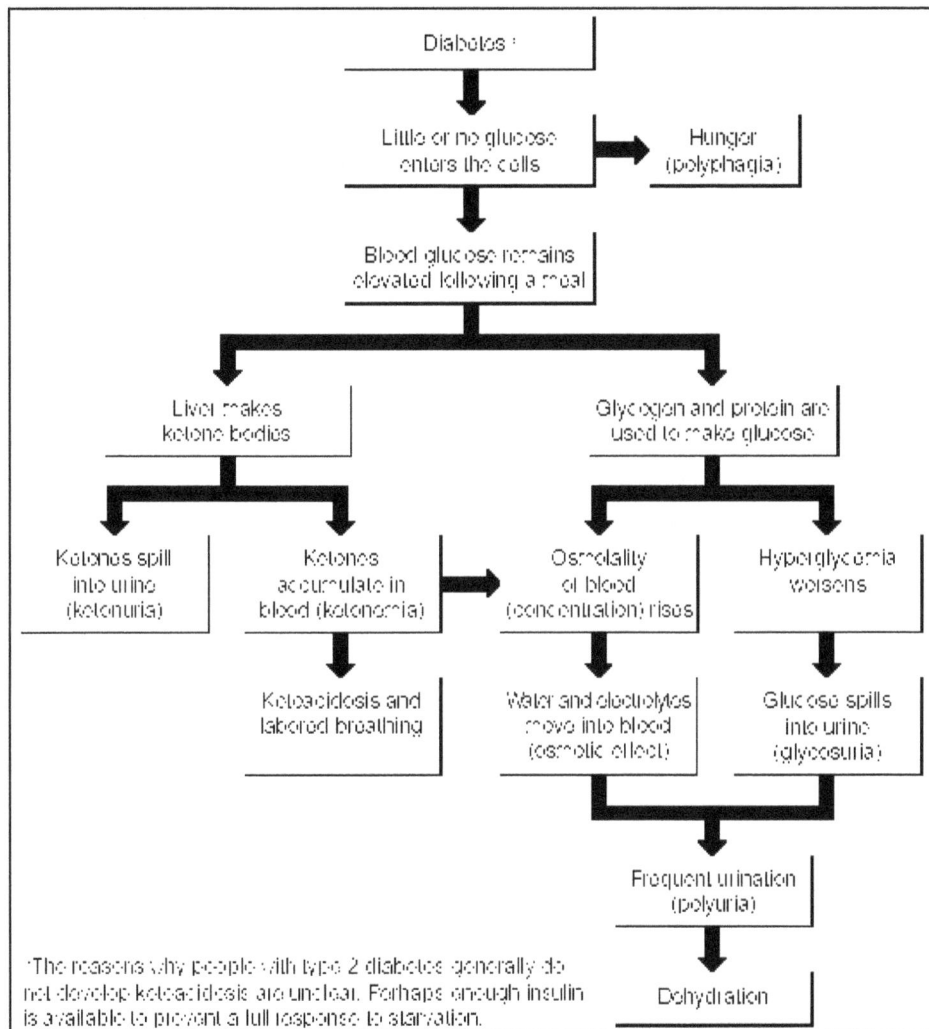

Figure 4.13: Summary of Diabetes.

☆ Maltitol: 2.1 kcal/g

☆ Mannitol: 1.6 kcal/g

☆ Sorbitol: 2.6 kcal/g

References

1. Shakuntala Manay, M. and Shadaksharaswamy, M., (1987). Foods-Facts and Principles, New Age International (P) Publishers Ltd., Chennai.

2. Wardlaw's, Perspectives in Nutrition, 8th Edition, McGraw-Hill Companies, ISBN 978–0–07–296999–3.

3. Nutrition by David C. Nieman, Diane E. Butterworth and Catherine N. Nieman, 1990 edition, Wm. C. Brown Publisher.

4. Bamji S. Mehtab *et al.,* Textbook of Human Nutrition, 3rd edition, 2010, Oxford and IBH Publishing Co. Pvt. Ltd., New Delhi. ISBN-978-81-204-1742-7.

5. TYPE 2 DIABETES AND DIET by NIN

6. Some factors affecting the digestion of glycemic carbohydrates and the blood glucose response, M Charlyn Vosloo, Journal of Family Ecology and Consumer Sciences, Vol. 33, 2005.

7. Gopalan.C, *et al.,* 2012, reprinted Nutritive Value of Indian Foods. NIN, Hyderabad

8. www.google.com

9. 2010 revised Recommended Dietary Allowances suggested by Indian Council of Medical Research.

$\boxed{\textbf{5}}$

Proteins

5.0 Introduction

Jons J.Berzelius, in 1838 named a group of chemical substances as proteins meaning of the first rank. The term protein was coined by dutch chemist G.T. Mulder in 1839 on the suggestion of Jons. J. Berzelius. The word protein was derived from the Greek word "proteios" which means "principal" or prime. These are the principal components of all living cells (of both plant and animal body) and are important in practically all aspects of cell structure and functions. The term protein signifies first or foremost and proteins are the most abundant macromolecules in cells and constitute over 50 per cent of dry weight of most of the organisms. Next to water, protein is the most abundant component of the body. It accounts for about 1/6th of the live body weight. Out of which a third of it is found in the muscles, a fifth in the bones and cartilage, a tenth in the skin and the remainder is in other tissues and body fluid. Proteins are large, complex molecules that play many critical roles in the body. These macromolecules are characterized by their nitrogen content. The nitrogen in proteins is in a specialform that can be readily used by our bodies for vital functions. In addition, some proteins contain small quantities of sulphur, phosphorus and minerals.

Plants have the ability to synthesize proteins from inorganic nitrogen sources such as ammonia, nitrate and nitrite. Animals unlike plants cannot synthesize proteins in this manner. Thus all animal life, either directly or indirectly, depends on plants to satisfy its daily protein requirement. Proteins are the instruments by which genetic information is expressed. They do most of the work in cells and are required for the structure, function, and regulation of the body's tissues and organs. Proteins are crucial to the regulation and maintenance of essential body functions. For example, maintenance of fluid balance, hormone and enzyme production, vision, and cell synthesis and repair each requires specific proteins. The body synthesizes proteins in many configurations and sizes, so that they can serve these greatly varied functions.

5.1 Compositions

Proteins are extremely complex nitrogenous organic compounds, built up by smaller units of structure called amino acids. They contain the elements- carbon, hydrogen, oxygen and nitrogen. Most of them contain sulphur and someproteins also contain phosphorous. In addition, some specialized protein too contains other trace elements like iron, iodine and copper.

The presence of nitrogen about 16 per cent distinguishes protein from carbohydrate and fat. Protein is more complex than fats and carbohydrates because of the large size of the molecules and great variation in the units from which it is formed *i.e.* the amino group.

5.2 Amino Acids

Proteins are made up of simpler compounds called amino acids. Amino acids can be proteinogenic amino acid and non proteinogenic amino acids. Proteinogenic amino acids are the amino acids that are precursors to proteins and are produced by cellular machinery coded for and in the genetic code of any organism. There are 22 standard amino acids, but only 21 are found in eukaryotes. Of the 22, selenocysteine and pyrrolysine are incorporated into proteins by distinctive biosynthetic mechanism. The other 20 are directly encoded by the universal genetic code. Human can synthesize 11 of these 20 from each other or from other molecules of intermediary metabolism. The other 9 must be consumed in the diet and are thus called as essential amino acids. The word proteinogenic amino acid means "protein building". Proteinogenic amino acid can be condensed into a polypeptide through a process called as Translation. In contrast non proteinogenic amino acid are either not incorporated in proteins (like GABA, L-DOPA) or are not produced directly and in isolation by standard cellular machinery. They are obtained from posttranslational modification of proteins.

The amino acids contain an amino (basic) and an acid (Carboxyl) group attached to the same carbon atom, the α - carbon atom in their molecules and in addition they also possess a third group denoted by "R". The structure of the amino acids may be represented as:

$$^+NH_3 \qquad\qquad NH_2$$
$$| \qquad\qquad\qquad |$$
$$R\text{-}C\text{-}COO^- \longleftrightarrow R\text{-}C\text{-}COOH$$
$$| \qquad\qquad\qquad |$$
$$H \qquad\qquad\qquad H$$

Zwitterion or Ionized   Unionized

Figure 5.1: Structure of a Amino Acid, R is Side Chain.

Many different amino acids are possible, by varying the group (R) that is attached to the carbon containing the amino group. Depending on the pH of the surrounding medium the amino acids can be in different ionic forms- acidic, zwitter ion or basic. All amino acids in the crystalline state or in aqueous solution at neutral pH values are ionized. The carboxyl group loses a proton and acquires a negative charge, hile the amino group gains a proton and hence acquire a positive charge. As a consequence, amino acids possess dipolar or zwitterion characteristics. At low pH, which is present in living cells a proton will be added on to the ionized carboxyl group producing a cation in which the α – amino group is ionized. On increasing the pH above neutrality, the amino group will give up a proton and the molecule will become an anion in which only the α – carboxyl group is ionized. This amphoteric nature of amino acids to act both as acid and as base is responsible for their buffering nature.

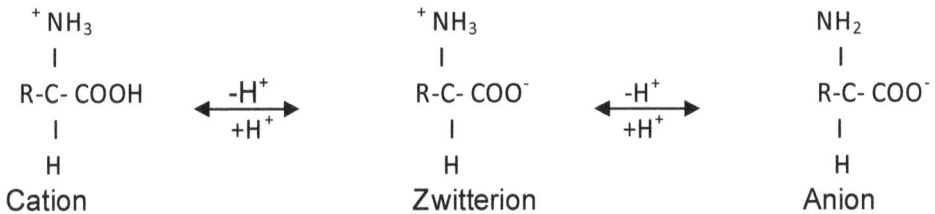

$$
\begin{array}{ccc}
\overset{+}{N}H_3 & \overset{+}{N}H_3 & NH_2 \\
| & | & | \\
R\text{-}C\text{-}COOH & R\text{-}C\text{-}COO^- & R\text{-}C\text{-}COO^- \\
| & | & | \\
H & H & H \\
\text{Cation} & \text{Zwitterion} & \text{Anion}
\end{array}
$$

$$\xrightleftharpoons[+H^+]{-H^+} \qquad \xrightleftharpoons[+H^+]{-H^+}$$

Amino acids also exhibit another important property namely optical isomerism. If the side chain R is other than H, and the alpha carbon atom (C next to the carboxyl group) becomes asymmetric with four different groups. Such compounds show optical isomerism *i.e.* they rotate the plane of polarized light. All amino acids except glycine, have a chiral centre. Depending on the configuration of atoms around the asymmetric carbon the amino acids are said to be in D or L configuration. In general the proteins from mammalian sources contain only L-amino acids. D-amino acids have however been identified in some peptide antibiotics of microbial origin. Amino acids with polar uncharged R groups are hydrophilic. Their side chains ionize at physiological pH.

Amino acids can be classified according to the polarity of R group, it follows as

Table 5.1: Amino Acid Classification According to Polarity of R Group

Group	Name of Amino Acid
Non polar Group	Alanine, Isoleucine, Leucine, Methionine, Phenylalanine, Proline, Tryptophan, Valine
Polar but uncharged	Asparagine, Cysteine, Glutamine, Glycine, Serine, Threonine, Tyrosine
Negatively Charged	Aspartic acid, Glutamic acid
Positively Charged	Arginine, Histidine, Lysine

Amino acids are also classified as Aliphatic- Glycine, Serine, Threonine, Leucine, Isoleucine and valine. Aromatic -Tryptophan, Tyrosine and Phenylalanine, Acidic- Aspartic acid, Glutamic acid, Basic-Arginine, Histidine, Lysine and sulphur containing amino acids- Cysteine, Methionine.

The body needs 20 different amino acids to function. All amino acids are important for proper functioning of the body, but 11 of these amino acids do not need to be acquired from the diet. They are classified as **non-essential (or dispensable)** amino acids because our bodies make them, using other amino acids that we consume. Our body cannot make the remaining 9 amino acids and these are essential (or indispensable) amino acids because they must be obtained from foods. **Essential amino acids** cannot be synthesized in the body because body cells cannot make the carbon skeleton of the amino acid, cannot attach an amino group to the carbon skeleton, or simply our body cannot perform the process fast enough to meet the body's needs.

Essential Amino Acids- Histidine, Isoleucine, Leucine, Lysine, Methionine, Phenylalanine, Threonine, Tryptophan, Valine.

Non-essential Amino Acids- Tyrosine, Serine, Proline, Glycine, Glutamine, Glutamic acid, Cysteine, Arginine, Asparagine, Aspartic acid, Alanine.

Few nonessential amino acids may be classified as "conditionally essential" or "semiessential" amino acids during infancy, disease, or trauma. For example, a person with the genetic disease phenylketonuria (PKU) has a limited ability to metabolize the essential amino acid phenylalanine due to a deficiency of the enzyme phenylalanine hydroxylase. This enzyme is needed to convert phenylalanine to the nonessential amino acid tyrosine. As a result, individuals with PKU cannot produce sufficient tyrosine, thereby making tyrosine a **"conditionally" essential amino acid** because it must be obtained from the diet. Following trauma and infection, the amino acids glutamine and arginine may be considered conditionally essential because supplemental amounts have been shown to promote recovery.

5.3 Peptide

Most peptides are the partial hydrolytic products of proteins. When the α-amino group of one amino acid reacts with the α-carboxyl group of another, a peptide bond is formed with the elimination of a molecule of water. Two amino acid molecules can be covalently linked through a substituted amide linkage termed as peptide bond to yield a dipeptide. When a few (2-10) amino acids are joined together by peptide bonds the compound is called an oligopeptide. With more than ten amino acid residues the compound is termed as a polypeptide. When the polypeptide contains about 100 amino acid units, it is called a protein. The linear peptide chain has two terminal residues, one terminal residue possessing a free amino group called the N-terminal residue, and the other a free carboxyl group called the C-terminal residue. The terminal residues are usually free to ionize. The sequence of amino acids determines each protein's unique 3-dimensional structure and its specific function. It is Amino acid sequences of different protein enable them to serve their different function. However a few peptides are of metabolic importance and are found free in nature. Carnosine and anserine are peptides found in muscle. Glutathione is found in mammalian erythrocytes and functions in oxidative metabolism. Oxytoxin and vasopressin are the examples of peptide hormones. Peptides give a colour reaction that is not given by free amino acids, the "biuret" reaction.

5.4 Classification of Proteins

An infinite number of proteins could be synthesized from the twenty off natural amino acids which have wide structural and functional diversity. It is difficult to classify the proteins on the basis of a single property or characteristic. However following are some common basis of classifying the Proteins:

5.4.1 Physical and Chemical Properties

This classification is sub-divided into further three groups according to the solubility.

5.4.1.1 Simple Proteins

Simple proteins are those which are made of amino acid units, each joined by a peptide bond. These will yield only amino acids upon hydrolysis by acids, alkalis or enzymes. Examples are albumins and globulins found within all body cells.

Table 5.2: Protein Classification

Sl.No.	Type	Examples
1.	Albumins	Egg albumin, serum albumin, lactalbumin
2.	Globulin	Tissue globulin, serum globulin
3.	Gliadin	Wheat gliadin, hordein (barley)
4.	Albuminoids	Keratin of hair, skin, Egg shell and bones, collagen and elastin of tendons, ligaments and bones.
5.	Histones	Globin of haemoglobin
6.	Protamine	Salmine from spermatozoa of salmon fish.

Classification of Simple Proteins by Solubility

After two German chemists, Emil Fischer and Franz Hofmeister, independently stated in 1902 that proteins are essentially polypeptides consisting of many amino acids, an attempt was made to classify proteins according to their chemical and physical properties, because the biological function of proteins had not yet been established. (The protein character of enzymes was not proved until the 1920s.) Proteins were classified primarily according to their solubility in a number of solvents. This classification is no longer satisfactory, however, because proteins of quite different structure and function sometimes have similar solubilities; conversely, proteins of the same function and similar structure sometimes have different solubilities. The terms associated with the old classification, however, are still widely used. They are defined in Table 5.3.

Albumins are proteins that are soluble in water and in water half-saturated with ammonium sulfate. On the other hand, globulins are salted out (*i.e.,* precipitated) by half-saturation with ammonium sulfate. Globulins that are soluble in salt-free water are called pseudoglobulins; those insoluble in salt-free water are euglobulins. the glutelins in acidified or alkaline solution.

Table 5.3: Protein Classification Based on Solubility

Name	Solubility in				Food Source
	Water	Salt Solution	Acid/Alkali	Alcohol 80 per cent	
Albumin	Soluble	Soluble	–	–	Milk, egg, plant and animal cells
Globulins	–	Soluble	–	–	Milk, egg, meat, plant cells, particularly in seed proteins.
Histones	Soluble	–	Soluble	–	Glandular tissues, pancreases, thymus, fish, cell nuclei,
Glutelins	–	–	Soluble	–	Cereal grains and related plant material
Prolamines	–	–	–	Soluble (50-80 per cent solution)	Cereal grain and related plant materials
Protamines	Soluble	–	Soluble	–	Fish sperm (80 per cent arginine)
Scleroproteins (albuminoids)	–	–	–	–	insoluble proteins of animal organs include keratin, hair, collagen

5.4.1.2 Conjugated Proteins

They are composed of simple proteins combined with a non-protein substance. The non protein substance is called as prosthetic group or cofactor. Following are few examples:

Table 5.4: Conjugated Protein Classification

Sl.No.	Type	Examples
1.	Glycoprotein	Ovomucoid of egg white containing a carbohydrate moiety.
2.	Lipoproteins	HDL (High Density Lipoproteins), LDL (Low Density Lipoproteins) and VLDL (Very Low Density Lipoproteins), These have lipids as prosthetic group.
3.	Chromoproteins	Haemoglobin which has iron-porphyrins as prosthetic group, carotenoids, bile pigments, and melanin
4.	Phosphoproteins	Casein in milk and vitellin egg yolk have phosphoric acid as prosthetic group.
5.	Nucleoproteins	Ribosomes and viruses contain nucleic acid as prosthetic group.
6.	Mucoproteins	Follicle stimulating hormones, β-ovomucoid
7.	Metalloproteins	Alcohol dehydrogenase a Zinc containing enzyme

5.4.1.3 Derived Proteins

These are not naturally occurring proteins and are the substances resulting from the decomposition of simple and conjugated proteins by the action of enzymes and chemical agents, heat, mechanical shaking, UV or X-rays. Following are the examples:

1. Primary such as myosin and fibrin etc.
2. Secondary such as peptones, peptides and proteoses etc.

5.4.2 According to the Amino Acid Structure

They can be Essential, non-essential and semiessential amino acids depending on the ability of the body to synthesize the amino acids. The extent to which the amino acids are present determines the quality of protein in any food.

5.4.3 According to Nutritional Qualities

Proteins may be broadly divided into three groups in regard to their nutrition value.

5.4.3.1 Complete Protein

These proteins maintain both life and support the growth. A complete protein contains essential amino acids to maintain body tissues and to promote a normal rate of growth and is referred as having a high biological value. Examples are egg, milk and meat (including poultry and fish) proteins, wheat germ and dried yeast have a biological value approaching that of animal source. This can include all animal proteins except gelatin and Plant protein-Gluten.

5.4.3.2 Partially Complete Proteins

They will maintain life, but lack sufficient amounts of some of the essential amino acids necessary for growth. Adults under no physiological stress can maintain satisfactory nutrition for indefinite period when consuming sufficient amount of protein from certain cereals or legumes-Gliadin of wheat, Prolamin of rye, Hordein of barley.

5.4.3.3 Totally Incomplete Proteins

These type of proteins are incapable of replacing or building new tissues and hence can neither support life nor can promote growth. Zein in corn is an examples of this type.Wheat germ is an incomplete protein because it is deficient in tryptophan because of this deficiency only less of the total protein can be used.

5.4.4 According to their Biological Function

Proteins are involved in different functions of the body and also makes it a basis for their classification.

5.4.4.1 Enzymes (Enzymoproteins)

Those proteins which are highly specialized in their function with catalytic activity are called enzymes. These proteins regulate almost all biological reactions going on inside all living cells. There are about 2000 different enzymes has been recognized; each capable of catalyzing a different kind of biochemical reaction. Without enzymes no life forms are possible. Examples are Kinases, dehydrogenases.

5.4.4.2 Transport Proteins

Transport proteins are those proteins which facilitate import of nutrients into cells or releases of toxic products into surrounding medium. Examples are Hemoglobin (which is a globular protein present in RBC of blood can binds with oxygen when blood passes though longs and distributes oxygen through out the body cells to affect cellular respiration), Blood plasma contains lipoprotein (which carries lipids from the liver to other organs).

Other kinds of transport proteins present in cell membrane called carrier proteins which are specially adopted to bind and transport glucose, amino acids and other nutrients across the membrane into the cells.

5.4.4.3 Nutrient and Storage Proteins

Storage proteins are those stored inside the cells or tissue as reserved food and can be mobilized at the time of nutrient requirement to provide energy.

The seeds of many store nutrient protein required for the growth of embryonic plant. These include proteins in wheat corn and rice stored in endosperm, ovalbumin (egg white) stored in egg. Casein in milk, Ferritin in animal tissue are the nutrient proteins, myoglobin.

5.4.4.4 Contractile or Motile Proteins

Some proteins provide cells, and organisms with the ability to contract to change the shape or to move about. These proteins includes Actin and myosin; which are present in form of filamentous protein in muscle cells for functioning in the contractile systems.

Tubulin is another contractile protein present in each cell type inform of micro tubes. Micro tubes are main constituent of cilice and flagella which help in movement of cells.

5.4.4.5 Structural Proteins

Many proteins give support and structure. They serve as a supporting filaments, cables or sheets to give biological structures strength and protection. This type of protein form major component of tendons, cartilages and bones. These include

☆ Fibrous proteins which are long and are less soluble in water. These physical properties are consistent with their biological role as structural proteins. Ligaments are contains special structural protein capable of stretching in two dimensions called as elastin. Hairs finger nails, feathers of birds consists of tough insoluble protein named keratin. Major component of silk fibers, threads of spider web contain structural protein named fibroin. Fibrin, elastin, keratin and collagen are included in the group.

☆ Globular Proteins are compact and spherical and usually are soluble in water. Enzymes, hormones, antibodies, transport proteins and respiratory proteins have a globular structure. Thus the globular proteins show more diverse functional properties because of their less rigid structures.

5.4.4.6 Defense Proteins

Many proteins in body of organisms posses defensive action against the invasion and attack of foreign entities or protect the body from injury. Among these proteins special globular protein named immunoglobulin's or antibodies in vertebrate's body is the most indispensible protein. It synthesized by lymphocytes and they can neutralize the foreign protein produced by bacteria, virus and other harmful microbes called antigens through precipitation or glutination.

Fibrinogen and thrombin are blood proteins belong to this class of proteins which help in blood clotting and protect blood loss from injury and help in repair.

5.4.4.7 Regulatory Proteins

Some proteins help to regulate or coordinate cellular or physiological activity. Among them are many hormones, such as insulin and glucagon; which are a regulatory protein formed in pancreatic tissue help to regulate the blood sugar level.

Growth hormones of pituitary and parathyroid hormones regulate Ca^{++} and phosphate transport in body. Other proteins called repressors regulate biosynthesis of enzymes.

5.4.4.8 Other Functional Proteins

There are number of proteins whose functions are not yet specified and are rather unusual. These includes:

☆ **Monelin:** A protein of an African plant has an intensely sweet taste and used as non toxic food sweetener for human use.

☆ **Antifreeeze:** A protein present in blood plasma of Antarctic fisher which protect their blood freezing in ice cold water.

☆ **Resillin:** A type of protein present in wing hinges of some insects with elastic properties.

5.4.5 According to their Precipitation Property

1. **Emulsoids:** are those proteins which stay in solution with lot of bound water and are not coagulated by mild acid.

2. **Suspensoids**: are those proteins that stay suspended due to the repulsion between charges. These are coagulated at their isoelectric point or by the charge neutralization.

Example: Milk is a good example of both as on mild acidification *i.e.* pH of 4.7-5.3 the casein or suspensoid separates out and the other proteins remain in the solution and can be separated by filtration. These are called as milk serum or whey proteins and contain a number of proteins like lactoalbumin and lactoglobulin etc.

5.5 Structure of Proteins

5.5.1 Protein Organization

There are four different levels of structure are found in proteins.

1. The primary structure of a protein is the linear sequence of amino acids in the polypeptide chain. This determines the protein's structure. The unique sequence of amino acids is reponsible for many of the fundamental properties of the protein for example: changing one amino acid residue in the β chain of haemoglobin which contain 574 amino acid residues can bring about profound changes in the biological properties. The sequence of amino acid also determines the seconday and tertiary structures of protein.

2. Secondary structure are the polypeptide chain having a specific shape stabilized by hydrogen and sulfur bonds. Amino acids must have accurate position in order for the amino acids to interact and fold correctly into the intended shape for the protein. This, in turn, allows chemical bonds (hydrogen and sulfur bonds) to form between amino acids near each other and stabilizes the structure.This creates a spiral-like shape called secondary structure. The seconday structure of the protein depends upon the structural characterisitcs of the peptide bond wich repeats itself along the chain.

When the restrictions of the peptide bon are superimposed in a poly amino acid chain of a globular protein, a α helix structure appears to be one of the most ordered and stable structures feasible. The α helix contains 3.6 amino acid residues per turn of the protein backbone wih the R-groups of the amino acids extending outward from the axis of the helical structure. Hydrogen bonding occurs between the nitrogen of one peptide bond and the oxygen of another peptide bond four residuess along the protein

Serine
Glycine
Alanine
Leucine
Valine
Lysine
Glycine

Primary ──→ **Secondary** ──→ **Tertiary** ──→ **Quaternary**

(a) Primary structure

Amino end

Carboxyl end

(b) Secondary structure

Hydrogen bonds between amino acids at different locations in polypeptide chain

α helix

Pleated sheet

(c) Tertiary structure

Heme

β polypeptide

(d) Quaternary structure

β   β

Heme group

α   α

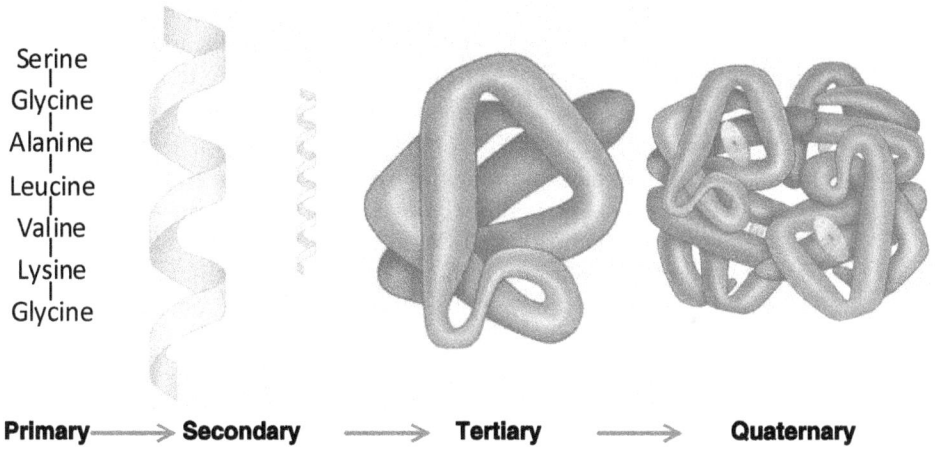

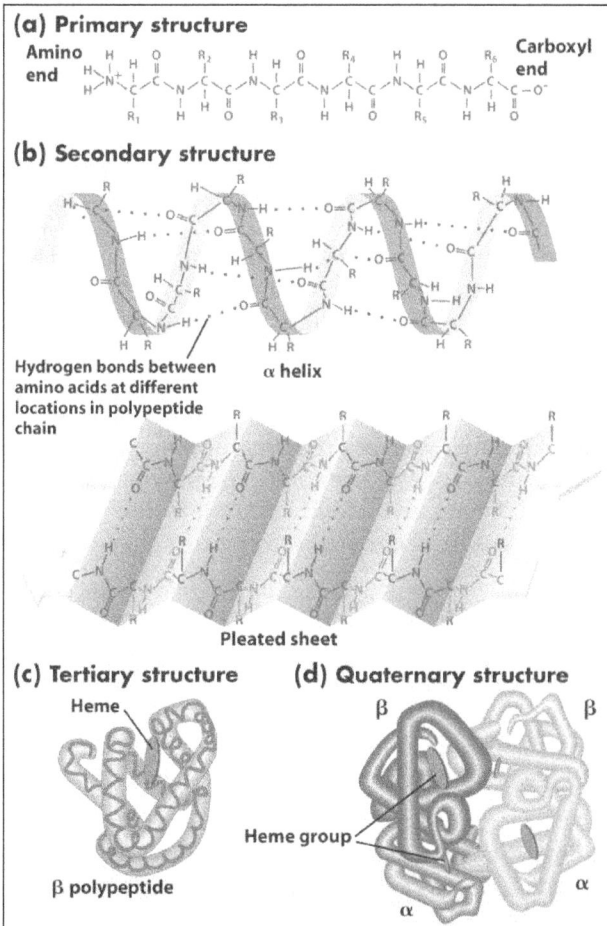

Figure 5.2: Protein Organization and Structure.
Source: **Perspectives in Nutrition, 8ᵗʰ edition, Gordon M. Wardlaw, Pal M. Insel.**

backbone structure. Another secondary structure found in many fibrous proteins is the β pleated sheet configuration. Ex are silk and insect fibres. Another type of secondary structure of fibrous proteins is the collagen helix.

3. Tertiary structure is the 3-dimensional shape of proteins. This structure involves the folding of regular units of the secondary structure as well as structuring of areas of the peptide chain devoid of secondary structure. The folded proteins are held together by hydrogen bonds formeed between R-groups, by electrostgatic interactions between chains possesing oppositely charged groups. It determines the physiological function of the protein. Thus, if a protein fails to form the proper configuration, then it cannot function.

   In the formation of tertiary structure, all the polar groups are on the surface of the molecule and the interior consists almost entirely of nonpolar hydrophobic residues such as those of leucine, valine, methionine and phenylalanine. The presence of polar R-groups on the surface of proteins usually accounts for their solubility in aqueous solutions.

4. Quaternary structure is one where 2 or more protein units join together to form a larger protein, such as hemoglobin. When a protein contains two or more polypeptide chains (subunits), the structure formed when individual polypeptide chains interact to form the native protein molecule, this is refered as quaternary structure.

Therefore, a protein may be active when the units are joined but inactive when the units are separate. There are three characteristics of assembled proteins which have influence on their diverse capabilities – each protein has a specific weight, definite amino acid sequence and specific three dimensional shape as it folds.

5.5.2 Denaturation of Proteins

Denaturation is the alteration of a protein's 3-dimensional structure of protein brought by exposure to acid or alkaline solutions, enzymes, heat, or agitation. This leaves the protein in a denatured state. Denaturation does not affect the protein's primary structure, but because of unraveling a protein's shape often affect the normal biological function.

In some instances, the denaturation of proteins is beneficial. For example, the secretion of hydrochloric acid in the stomach during digestion denatures food proteins, which helps increase their exposure to digestive enzymes and aids in the breakdown of polypeptide chains. The heat produced during cooking can also denature proteins, making them safer to eat (*e.g.*, pathogenic bacterial protein denaturation) and more pleasing to eat (*e.g.*, eggs solidify in cooking).

However, denaturation also can be harmful to physiological function and overall health. During illness, changes in gastrointestinal acidity, body temperature, or body pH can cause essential proteins to denature and lose their function.

5.6 Functions of Proteins

Proteins play their role in several crucial ways in body metabolism and in the formation of body structures. An individual receive the required amino acids for the synthesis of proteins from the diet as well as by the recycling of body protein. However, only when we eat adequateamount of dietary carbohydrate and fat, the dietary proteins can be used efficiently for these functions. So, in case we don't consume sufficientamount of carbohydrates and fats required to meet our energy needs, some amino acids will be used to produce energy, making them unavailable to build body proteins for other essential functions.

1. Transporting Nutrients

Many proteins function as transporters for other nutrients, carrying them through the bloodstream to cells and across cell membranes to sites of action. For example, Lipoproteins transport large lipid molecules from the small intestine, through the lymph and blood to body cells, hemoglobin carries oxygen from the lungs to cells.

Few minerals and vitamins also have protein carriers that aid in their transport into andout of tissues and storage proteins. This includes transferrin and ferritin (carrier and storage proteins for iron), ceruloplasmin (a carrier protein for copper), and retinol-binding protein (a carrier protein for vitamin A).

2. Producing Vital Body Structures

During periods of growth, new proteins are synthesized to support the development of vital body tissues and structures. The primary functions of protein is to give structural support to body cells and tissues. Each cell in the body contains proteins; in the cell membrane that surrounds the cell and holds it together; in the organelles in the cytoplasm and in the nucleus itself where the DNA is housed. The key structural proteins are, actin, myosin and collagen which constitute more than a third of body protein and provide a matrix for muscle, connective tissue, and bone. So, the deficiency of proteins can results in many diseases like kwashiorkar.

3. Maintaining Fluid Balance

The blood proteins albumin and globulin are important in maintaining fluid balance between the blood and the surrounding tissue space. Normal blood pressure in the arteries forces blood into capillary beds. The blood fluid then moves from the capillary beds into the spaces betweennearby cells (interstitial spaces) to provide nutrients to those cells.Proteins such as albumin are too large to move out of the capillary beds into the tissues. The presence of these proteins in the capillary beds attracts the right amount of fluid back to theblood, partially counteracting the force of blood pressure to maintain fluid balance. So, the plasma proteins (large molecules in the blood vessels), influence the movement of water from one area to another by exerting pressure, thus helping to maintain blood flow and circulation.

Proteins are also indirectly involved in the fluid balance *i.e.* movement of electrolytes.When protein consumption is inadequate, the concentration of proteins eventually decreasesin the bloodstream. Excessive fluid then builds up in the surrounding tissues becausethe counteracting force produced by the smaller amount of blood proteins is too weak topull enough of the fluid back from the tissues into the bloodstream. As fluid builds up in the interstitial spaces, the tissues swell, resulting in medicalcondition of protein deficiency edema.

4. Contributing to Acid-Base Balance

Proteins play animportant role in regulating acid-base balance and body pH (refer 5.2). For example, proteins locatedin cell membranes pump chemical ions into and out of cells. The ion concentrations thatresult from the pumping action help keep the blood slightly alkaline (pH = 7.35–7.45). Inthis way, proteins act as buffers—compounds that help maintain acid-base balance within anarrow range. Proteins are especially good buffers for the body because they have negativecharges, which attract positively charged hydrogen ions. This allows them to accept and release hydrogen ions as needed to prevent harmful changes in pH.

Acid is a byproduct of several chemical reactions that occur in the body. Normal cell functioning, such as enzyme activity, can occur only when the pH remains within a small range. Excess acid can disrupt normal body functioning and in severe cases result in coma or death. Proteins buffer this acid until it can be removed from the body by the kidneys or lungs.

5. Forming Hormones, Enzymes, and Neurotransmitters

Hormones act as messengers inthe body and aid in regulatory functions, such as controlling the metabolic rate and theamount of glucose taken up from the bloodstream. Amino acids are required for the synthesis of most hormones in the body. Some hormones,such as the thyroid hormones, are made from only 1 amino acid, whereas others,such as insulin, are composed of many amino acids.

Cells contain thousands of enzymes that facilitate chemical reactions fundamental to metabolism. Amino acids also are required for thesynthesis of enzymes. Many neurotransmitters (released by nerve endings), dopamine and norepinephrine (synthesized from the amino acid tyrosine), and serotonin (synthesized from the amino acid tryptophan).

6. Contributing to Immune Function

Antibodies areproteins and are key component of the immune system. Antibodies can bind toforeign proteins (called antigens) that attack the body and can avert their attack ontarget cells.

In a normal, healthy individual, antibodies are very efficient in combatingthese antigens to prevent infection and disease. However, without sufficient dietary protein,the immune system lacks the substance needed to build this defense. Thus, immuneincompetence (called anergy) develops and reduces the body's ability to fight infection.

7. Forming Glucose by Gluconeogenesis

The body has to maintain a constant concentration of blood glucose to supply energy,especially for red blood cells, brain cells, and other nervous tissue cells that dependon glucose for energy. If carbohydrate intake is inadequate to maintain blood glucose levels, the liver (and kidneys, to a lesser extent) is forced to make glucosefrom the amino acids present in body tissues. This process is called gluconeogenesis.

Making glucose from amino acids results in the development of widespread muscle wasting inthe body (called **cachexia**).

8. Providing Energy

Under normal conditions, body cells use primarily fats and carbohydrates for energy. Proteins supply very little energy for healthy individuals. The amino acids which are not used for protein synthesis are broken down to provide energy, 1g of protein gives rise to 4.2kcal.

On average proteins and carbohydrates provide 4 kcal/g but utilizing proteins for energy generation is very costly source of energy because of the amount of metabolism and processing the liver and kidneys must have to perform (23 per cent) to use proteins as energy source. When carbohydrate is not sufficient to meet the glucose needs for central nervous system, then some amount of protein will be converted to supply glucose or if it is above the caloric requirements then it will be metabolized into fat.

5.7 Sources of Proteins

Plants are the major source of protein in many areas of the world. Globally in a normal person diet about 35 per cent of protein comesfrom animal sources. The best source of animal protein for growing children is milk. Milk also provides good amount of calcium in the diet of vegetarian. Egg, Fish are source of good quality protein. Skimmed milk is a rich source of protein as whole milk. Butter milk of good quality can also serve as a source of good quality proteins.

Protein is supplied by the diet, as well as by the recycling of body protein. Dietary protein is needed to replenish and maintain an adequate amino acid pool for protein synthesis and repair.

Our body can recycle the amino acids and to the amino acid pool. The amino acids from digested proteins are absorbed rapidly into the blood and passed onto different tissues to meet their needs. Some non-essential amino acids are synthesized in the liver and also released into the circulation. The amino acids released by hydrolysis of tissue proteins, alsothe intestinal tract lining is constantly sloughed off and the digestive tract treats sloughed cells just like food particles and absorbs their amino acids released during digestion. Amino acids so released are then added to the amino acid pool in the body. Thus the protein metabolism of mammals is in a dynamic state and the synthesis and breakdown of tissue protein takes place constantly. The unwanted amino acids are oxidized in the liver to yield energy and urea. The dynamic aspects of protein metabolism are represented in Figure 5.3

Plants can provide ample amounts of dietary protein in addition to providing fiber and a variety of vitamins, minerals, and phytochemicals. Plant proteins also contain no cholesterol and little saturated fat, unless added during processing. So the peoples can be benefitedby adding soy and other plant proteins to their diets foras a source of dietary protein with decreased risks of cardiovascular disease, certain cancers, obesity, and diabetes.

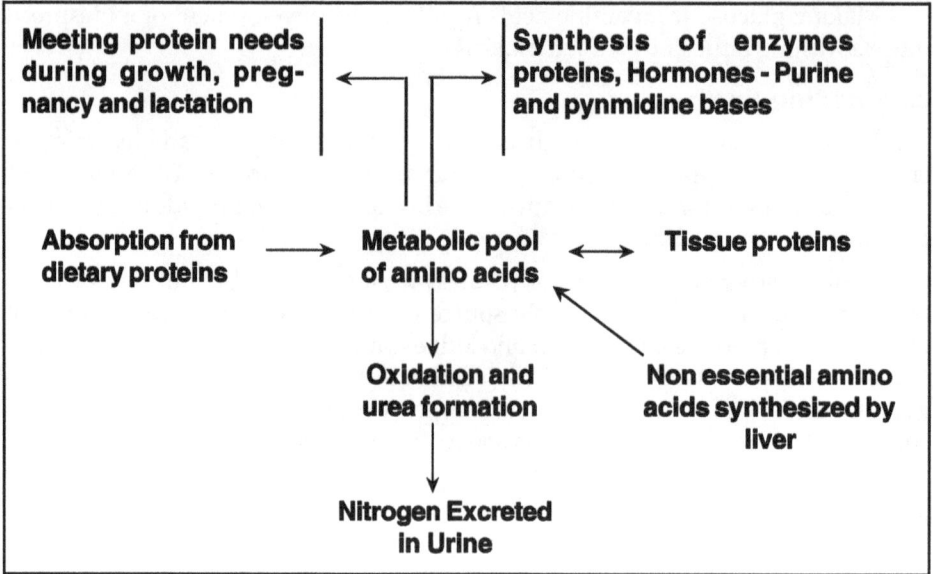

Figure 5.3: Dynamic Aspects of Protein Metabolism.

Potato contain good quality protein but the amount is less. Proteins in legumes/pulses are mostly concentrated in the aleurone layer and in cereals is found in embryo and aleurone layer.

5.7.1 Plant Sources

Table 5.5: Plant Sources of Protein

Group	Name	Protein (per cent)
Oilseeds	Soyabean	43.2
	Ground nut	26.7
	Cottonseed	19.5
	Sesame	18.3
	Sunflower seed	12.5
Pulses and Legumes	Lentils	25.1
	Black gram dal	24.6
	Green gram dal	24.5
	Bengal Gram dal	20.8
Fresh Vegetables	Fresh Peas	6
	Broad bean	4.5
	Jack fruit, tender	2.6

Source: RDA, From Nutritive Value of Indian Foods, 2012 ed.

5.7.2 Animal Sources

Proteins are found in animal tissues like blood, offals, muscle. The dietary sources are listed below:

Table 5.6: Animal Sources of Protein

Name	Protein (per cent)
Meat	18-22
Milk	3.5
Egg white	12
Fresh water fish	13-25

5.7.3 Microbial Sources

The proteins can also be obtained form microbial sources like algae, fungi, bacteria, yeasts etc. These are termed as Single Cell Proteins (SCP) and these are isolated form microorganisms.

5.8 Recommended Intakes of Protein

Proteins are required for maintenance (replacing the wear and tear in tissue) in adults, for growth in infants and children, for fetal development in pregnancy and lactation.

5.8.1 Based on Nitrogen

Healthy individuals who are not in periods of growth or recovering from illness or injury require to consume protein in an amount that replaces the protein lost in urine (primarily as urea), feces, sweat, skin cells, hair, and nails. So, a positive protein balance is to be maintained. There could be following protein equilibrium in the body as explained below:

1. **Equilibrium Condition**— Protein intake =Protein losses,

   Protein balance (or equilibrium) is maintained as long as energy intake is adequate to prevent the use of protein for energy.

2. **Negative protein (or nitrogen) balance**— Protein intake < Protein losses.

   Negative protein balance often develops in individuals eating inadequate dietary protein accompanied by a serious, untreated illness or injury and inthose with diseases (*e.g.,* Cushing's disease) that elevate the production of the hormonecortisol, which increases protein breakdown. Over time, the increased rates of protein breakdown will result in wasting of lean body mass. Negativeprotein balance eventually causes blood proteins, skeletal muscles, the heart, the liver, andother organs to decrease in size or volume. Only the brain resists protein breakdown.

3. **Positive protein (or nitrogen) balance**— Protein intake > Protein losses.

   During periods of growth and recovery from injury, trauma, or illness, positive protein balance is required to supply sufficient materials for building and repairing tissues. Merely eating more protein does not build additional body protein, the positive balance also requires an appropriate hormonal state. For example, the hormones insulin, growth hormone, and testosterone all stimulate protein synthesis for the building of new tissue. Howeverprotein needs must be increased above normal needs for this to occur.

Nitrogen is a constituent of protein and can be more easily measured;Researchers can measure dietary protein intake and body losses of protein to determine protein balance. So, for estimating the protein; nitrogen, rather than protein, is measured. As nitrogen makes up approximately 16 per cent of the weight of an amino acid (100/16= 6.25). Therefore, nitrogen intake multiplied by 6.25 provides an estimate of protein intake:

Protein (g) = Nitrogen (g) X 6.25

or

Nitrogen intake or output divided by 0.16 (16/100) yields a rough estimate of protein intake or output.

Nitrogen balance studies are difficult to conduct because an accurate measure of allsources of nitrogen intake and loss is needed over a 24-hour period which is not usually feasible. Thus, it is easier to calculate proteinneeds based on the RDA.

5.8.2 Based on RDA

The actual amount of protein to be consumed daily to meet the above requirement will depend upon the quality of dietary protein. The higher the quality, lower the requirement and vice-versa. The requirements are generally determined in terms of egg and for their proteins are computed after making adjustment for the lower quality of dietary protein relative to egg. The adult requirement of egg protein is 0.7g per kg body weight/day while requirement in terms of mixed vegetable protein is 1.0 g/kg body weight/day. The Indian Council of Medical Researchrecommends **1.0 g/kg/day** as the safe level of intake in terms ofdietary protein for Indians.

During pregnancy and lactation additional allowancesare recommended. Protein requirements for children varydepending on body weight and expected weight gain.The children require more protein per unit body weight than do adults, because of new tissues which are being laid down during growth are drawn from dietary proteins. Thus, a young child of 1-2 years require 1.2g egg protein/kg body weight/day or 2.0 g mixed vegetable protein/kg body weight/day. Similarly the protein requirement of women during pregnancy and lactation are more than normal state. However these values are valid only when other nutrients particularly the calories in diet are adequate.

Recommended Dietary Allowances for Indians

The RDA for protein does not address the additional protein amounts needed during recovery from illness or injury or that might be needed to support the needs of highly trained athletes. Protein needs range from approximately 0.8 to 2.0 g/kg body weight in recovery states and 0.8 to 1.7 g/kg body weight in endurance or strength athletes. Mental stress, physical labor, and routine weekend sports activities do not require an increase in the protein RDA.

Table: 5.7 RDA for Protein

Group	Physical Activity Group	Body Weight (kg)	Energy (kcal/day)	Protein (g)
Man	Sedentary, Moderate, Heavy work	60	2425, 2875, 3800	60
Woman	Sedentary, Moderate, Heavy work	50	1875, 2225, 2925	50
	Pregnant woman	50	+300	+15
	Lactating woman 0-6 months	50	+550	+25
	Lactating woman 6-12 months	50	+400	+18
Infants	0-6months	5.4	108/kg	2.05/kg
Boys	13-15 years	47.8	2450	70
Girls	16-18 years	49.9	2060	63

Source. RDA, From Nutritive Value of Indian Foods, 2012 ed.

5.9 Evaluation of Food Protein Quality

Various measures can be used to evaluate the protein quality of a food. These measures indicate a food protein's ability to support body growth and maintenance. Protein quality is determined primarily by the food's digestibility (amount of amino acids absorbed) and amino acid composition, compared with a reference protein (*e.g.*, egg white protein)known to provide the essential amino acids in amounts needed to support growth. Thedigestibility of animal proteins is relatively high (90–100 per cent), in contrast to that of plantproteins (70 per cent).

The concept of protein quality applies only under conditionsin which protein intakes are equal to or less than the amount of protein neededto meet the requirement for essential amino acids. When protein intake exceeds this amount, the efficiency of protein use is decreased, even with the highest-quality proteins.This occurs because, once essential amino acid needs have been met, the remainingamino acids (both essential and non-essential) cannot be readily stored and are primarilydegraded for use as energy. Following are the methods by which food protein quality can be determined:

5.9.1 Biological Value (BV)

The biological value (BV) of a protein is a measure of how efficiently the absorbed foodprotein is converted into body tissue protein. If a food contains adequate amounts ofall 9 essential amino acids, it should allow a person to efficiently incorporate amino acidsfrom food protein into body protein.

To determine the BV, nitrogen retention in the body is compared with the nitrogencontent of the food protein. More nitrogen is retained when a food's amino acid patternclosely matches the amino acid pattern of body protein. The better the match, the higherthe BV. In contrast, if the amino acid pattern in a food is quite unlike body tissue amino acid patterns then more amount of nitrogen is excreted because many of the amino acids in the food willnot be incorporated into body protein. The BV of such a food protein is low, as little ofthe nitrogen is retained in body tissues.

$$BV = \frac{\text{Nitrogen retained (g)}}{\text{Nitrogen absorbed (g)}} \times 100$$

Egg white protein has a BV of 100, the highest BV of any single food protein. Thismeans that basically all nitrogen absorbed from egg protein is retained and incorporatedinto body tissue protein. Most animal proteins have a high BV, reflecting a tissueamino acid composition similar to that of human tissues. Plants have amino acid patternsthat differ greatly from those of humans. Therefore, the BV of plant proteins is usuallymuch lower than that of animal proteins.

5.9.2 Protein Efficiency Ratio (PER)

Protein efficiency ratio (PER) is another procedure for assessing a food's protein quality. The PER compares the amount of weight gain by a growing laboratory animal consuming a standardized amount of the protein being studied with the weight gain in an animal consuming a standardized amount of a reference protein, (casein- milk protein).

$$PER = \frac{\text{Weight gain (g)}}{\text{Protein consumed}}$$

The PER of a food reflects its biological value because the weight gain and growth measured in the PER are dependent on the assimilation of food protein into body tissue. Thus, animal proteins with a high BV also yield a high PER, whereas plant proteins generally yield a lower BV and PER because they are incomplete proteins. The FDA uses this method to set standards for the labeling of foods intended for infants.

5.9.3 Chemical Score

The protein quality of a food also can be evaluated by its chemical score. To calculate a Protein chemical score, the amount of each essential amino acid in a gram of the food protein being tested is divided by the "ideal" amount for that amino acid in a gram of the reference protein (usually egg protein). The lowest (or limiting) amino acid ratio that is calculated for the essential amino acids of the test protein is the chemical score of that protein. Chemical scores range from 0 to 1.0.

$$\text{Chemical score} = \frac{\text{mg of limiting amino acid per g of protein}}{\text{mg of limiting amino acid per g of an "ideal" protein}}$$

For example: Assume that the ideal lysine amount in a diet is 5.1 per cent or 5.1mg/100mg of total protein. Wheat protein is most deficient in lysine, it contain 2.4 per cent of total protein. So the chemical score for wheat is 2.4/5.1 = 0.47 or 47 per cent of needs.

5.9.4 Net Protein Utilization (NPU)

NPU is defined as Nitrogen retained to nitrogen intake multiplied by 100. This include the digestive losses of nitrogen also. NPU is generally determined at or below the nitrogen intake required for maintenance. In humans, it is calculated from nitrogen balance studies.

$$NPU = \frac{Nitrogen\ retained}{Nitrogen\ intake} \times 100$$

Table 5.7: Nutritive Value of Proteins

Protein	Biological Value (BV)	Net Protein Utilization (NPU)	Protein Efficiency Ratio (PER)
Animal Protein			
Egg	96	96	3.8
Milk	90	85	2.8
Meat	74	76	3.2
Fish	80	74	3.5
Vegetable Protein			
Rice	80	77	1.7
Cereals			
Wheat	66	61	1.3
Maize	50	48	1.0
Pulses			
Bengal Gram	74	61	1.1
Red Gram	72	54	1.7
Oilseeds			
Groundnut	55		

Source: Nutritive Value of Indian Foods, by C. Gopalan *et al.,* ed. 2012.

5.9.5 Amino Acid Score

$$Amino\ Acid\ Score = \frac{mg\ amino\ acid\ per\ g\ of\ test\ protein}{mg\ of\ amino\ acid\ per\ g\ of\ reference\ protein} \times 100$$

5.9.6 Protein Digestibility Corrected Amino Acid Score (PDCAAS)

The most widely used measure of protein quality is called the Protein Digestibility Corrected Amino Acid Score (PDCAAS). This score is obtained by multiplying a food's chemical score by its digestibility. For example, to estimate the PDCAAS of wheat, its chemical score (0.47) is multiplied by its digestibility (0.90). This gives a PDCAAS of approximately 0.40. The highest PDCAAS is 1.0, which is the score for soy protein and most animal proteins. Thus a protein which doesn't contain any of the 9 essential amino acids (*e.g.*, gelatin) has a PDCAAS of 0 because its chemical score is 0.

PDCAAS = Chemical score × Digestibility

PDCAAS of wheat is 0.40, Other PDCAAS values are egg white, 1.0; soy protein, 0.92 to 0.99; beef, 0.92; and black beans, 0.53.

5.10 Protein Digestion and Absorption

For some foods, the first step in protein breakdown takes place during cooking. Cooking unfolds (denatures) proteins and softens the tough connective tissues in meat. This can make many protein-rich foods easier to chew and aids in breakdown during digestion and absorption in the GI tract.

Our body will stimulates gastrin-producing cells in the stomach to release the hormone (gastrin) as an individual think about food or eating food. Gastrin also strongly stimulates the stomach's parietal cells to produce acid, which aids in digestion and the activation of pepsin. Pepsin is actually stored as an inactive enzyme (called pepsinogen) to prevent it from digesting the stomach lining. Once pepsinogen enters the stomach's acidic environment (pH between 1 and 2), part of the molecule is split off, forming the active enzyme pepsin. The hormone gastrin controls the release of pepsin.

The enzymatic digestion of protein begins in the stomach with the secretion of hydrochloric acid. Once proteins are denatured by stomach acid, pepsin, a major enzyme produced by the stomach for digesting proteins, begins to break the long polypeptide chains into shorter chains of amino acids through hydrolysis reactions. Pepsin does not completely break up proteins into amino acids as it can break only a few of the many peptide bonds found in these large molecules.

From the stomach, the partly digested proteins move with the rest of the nutrients and other substances present in a meal (called as chyme) into the duodenum. Once in the small intestine, the polypeptide units, and any lipid associated with it will cause the release of the hormone cholecystokinin (CCK) from the walls of the small intestine. CCK, will then stimulates the pancreas to release the protease or protein-splitting enzymes trypsin, chymotrypsin, and carboxypeptidase into the small intestine. These enzymes will act on the polypeptides to give short peptides and amino acids. The short peptides and amino acids (in the lumen of the small intestine) are actively absorbed into the cells of the small intestine (~11 different transport mechanisms for amino acids have been identified within the absorptive cells of the small intestine).

The left short peptides are then broken down to individual amino acids by peptidase enzymes. Amino acids then travel via the portal vein to the liver for use in protein synthesis, energy requirements, conversion to carbohydrate or fat, or release into the bloodstream for transport to other cells. Intact proteins cannot be absorbed from the digestive tract. Except during infancy (up to 4 to 5 months of age), the gastrointestinal tract is fairly permeable to small proteins, so few whole proteins can be absorbed. Because proteins from foods such as cow's milk and egg white may predispose an infant to food allergies, pediatricians recommend waiting until an infant is a year old in age or older before introducing common allergenic foods.

Mentioned below is the systematic protein digestion process:

How the enzymes act on the peptide linkage?

There are two types of enzymes which breaks the linkage, they are endopeptidases and exopeptidases.

1. Action of Endopeptidases will result in in large polypeptides, oligopeptides, and free amino acids. Endopeptidases will act on the linkage between amino acids.

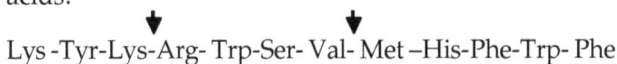

   Lys -Tyr-Lys-Arg- Trp-Ser- Val- Met –His-Phe-Trp- Phe

2. Action of exopeptidases which will digests external peptide bonds and the products of exopeptidase digestion amino acids only.

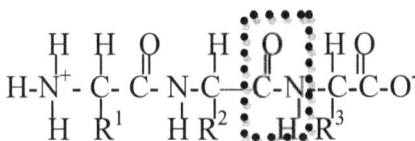

$$\underset{\text{Amino end}}{} \quad \text{H-N}^{+}\text{- C- C- N-C-C-N-C-C-O}^{-} \quad \underset{\text{Carboxyl end}}{}$$

Amino end   Carboxyl end

1. Protein Digestion-Mouth

   ✰ No protein digestion occurs in the mouth.

2. Digestion in the Stomach

Protein digestions begins in the stomach by gastric juice. HCl is released from parietal cells in response to gastrin.

a. Role of Gastric HCl

   ✰ It causes denaturation of proteins (2, 3, and 4 structure).

   ✰ It converts proteins to metaproteins, which are easily digested.

   ✰ It converts pepsinogen(inactive) to pepsin(active) by removing 46 amino acids to activate pepsin.

   ✰ It makes pH in the stomach suitable for the action of pepsin.

b. Pepsin

   ✰ It is an endopeptidase acting on central peptide bond in which amino group belongs to aromatic amino acids *e.g.* phenyl alanine, tyrosine and tryptophan.

   ✰ It is secreted in an active form called pepsinogen.

   ✰ Its optimum pH is 1.5- 2.2.

   ✰ It is activated by HCl then by autoactivation.

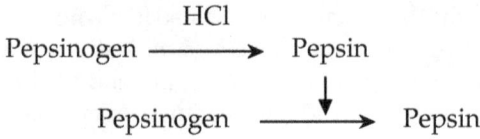

$$Pepsinogen \xrightarrow{\text{HCl}} Pepsin$$

$$Pepsinogen \xrightarrow{\qquad} Pepsin$$

c. Rennin

- ☆ It is a milk clotting enzyme.
- ☆ It is present in stomach of infants and young animals.
- ☆ Its optimum pH is 4.
- ☆ It acts on casein converting it to soluble paracasein, which in turn binds calcium ions forming insoluble calcium paracaseinate. Calcium paracaseinate is then digested by pepsin.

$$Casein \xrightarrow{\qquad} Paracasein \xrightarrow{\qquad} Calcium\ paracaseinate$$

d. Gelatinase

- ☆ It is an enzyme which liquefies gelatin.

  The end products of protein digestion in the stomach are proteoses, peptones and large polypeptides.

3. Digestion in the Small Intestine

Digestion of proteins is completed in the small intestine by proteolytic enzymes present in pancreatic and intestinal juices. With the release of chyme into SI it stimulates the release of secretin and CCK (cholecystokinin). Then secretin and CCK stimulate release of bicarbonate, water, electrolytes, and zymogens from pancreas.

A. Pancreatic Juice

1. Trypsin

- ☆ It is an endopeptidase with an optimum pH of 8.0. This peptidase hydrolyzes central peptide bond in which the carboxyl group belongs to basic amino acids eg. Arginine, histidine and lysine.
- ☆ It is secreted in an inactive form called as trypsinogen. And is activated by enterokinase enzyme then by autoactivation.

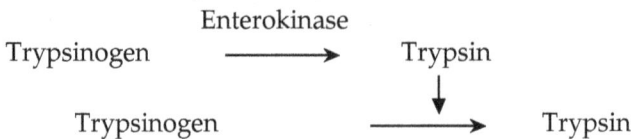

$$Trypsinogen \xrightarrow{\text{Enterokinase}} Trypsin$$

$$Trypsinogen \xrightarrow{\qquad} Trypsin$$

2. Chymotrypsin

- ☆ It is an endopeptidase having optimum pH of 8.0. This peptidase hydrolyzes central peptide bond in which the carboxyl group belongs to aromatic amino acids.
- ☆ It is secreted in an inactive form called as chymotrypsinogen and is activated by trypsin.

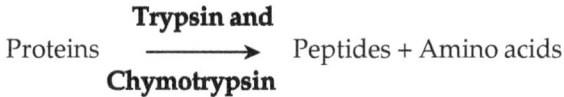

$$\text{Proteins} \xrightarrow{\textbf{Trypsin and Chymotrypsin}} \text{Peptides} + \text{Amino acids}$$

3. Elastase

☆ It is an endopeptidase have an optimum of pH 8.0, it primarily acts on peptide bonds formed by serine, alanine and glycine. It is secreted in an inactive form called proelastase and is also activated by trypsin.

☆ It digests elastin and collagen.

4. Carboxypeptidase

☆ It is an exopeptidase with an optimum pH of 7.4. This peptidase hydrolyses the terminal (peripheral) peptide bond at the carboxyl terminus (end) of the polypeptide chain.

☆ It is secreted in an inactive form called as procarboxypeptidase and is activated by trypsin

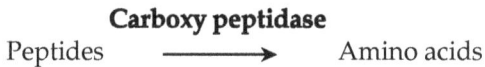

$$\text{Peptides} \xrightarrow{\textbf{Carboxy peptidase}} \text{Amino acids}$$

B. Intestinal Juice

Intestinal juices contains amino peptidase (polypeptidases and dipeptidases) which hydrolyse polypeptide and dipeptide to individual amino acids.

1. Aminopeptidase

It is an exopeptidase that acts on the terminal peptide bond at the amino terminus of the polypeptide chain and it releases a single amino acid.

There are several peptidases acting on different proteins.

2. Dipeptidase

☆ It acts on dipeptides and releases 2 amino acids.

☆ The end products of protein digestion in the small intestine are amino acids.

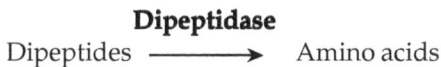

$$\text{Dipeptides} \xrightarrow{\textbf{Dipeptidase}} \text{Amino acids}$$

3. Tripeptidase

It acts on tripeptides and it release a single amino acid and dipeptide.

Tripeptides **Tripeptidase** Amino acids

Proteins are mainly absorbed in the form of amino acids. Amino acids are absorbed by active transport mechanism in the intestinal cells. Sometimes whole protein may be absorbed by the mechanism of **pinocytosis**.

4. Digestion in the Large Intestine

☆ No digestion occurs here.

☆ When undigested protein enter large intestines, bacteria causes nitrification of proteins leading to foul smelling flatus.

☆ Absorbed amino acids pass into the portal blood and reach liver where they are converted to proteins. Other amino acids are transported through general circulation and are utilized for protein synthesis in the tissues.

Protein Absorption

It is an active process that require energy. Required energy is derived from hydrolysis of ATP. This process occurs in small intestine and the absorption of amino acids is rigid in the duodenum and jejunum but is slow in the ileum.

Peptide Absorption

☆ Dipeptides and tripeptides are absorbed via active transport using PEPT1 transport protein.

☆ Peptide absorption accounts for 60 per cent of proteins absorbed.

☆ Peptide absorption occurs faster than free AA absorption.

Step 1: Hydrogen binds to PEPT1.

Step 2: Di- or tripeptide binds to PEPT1.

Step 3: PEPT1 releases hydrogen and di-or tripeptide inside enterocyte.

Step 4: Hydrogen is pumped back into the SI lumen in exchange for sodium.

Step 5: Sodium is pumped out of enterocyte and potassium is pumped in using ATP.

Step 6: Peptidase(an enzyme in the cytoplasm) digests di- and tripeptides to free AA.

Mechanics of Amino Acids Absorption

There are two mechanisms for amino acids absorption.

1. Carrier Protein Transport System

- It is main system for amino acid absorption and is an active process which require energy. The required energy is derived from ATP. Absorption of one amino acid molecule requires one ATP molecule. There are 7 carrier proteins, one for each group of amino acids. Each carrier proteins has two sites; one for amino group and one for Na^+. It co-transports amino acids and Na^+ from intestinal lumen to cytosol of intestinal mucosa cells. The absorbed amino acid passes to the portal circulation, while Na^+ is extruded out of the cell in exchange with K^+ by sodium pump.

2. Glutathione Transport System

Glutathione is used to transport amino acids from intestinal lumen to cytosol of intestinal mucosa cells.It is also an active process that needs energy. The energy required is derived from ATP. Absorption of one amino acid molecule require 3 ATP molecules.

☆ Glutathione reacts with amino acid in the presence of γ-glutamyl transpeptidase to form γ-glutamyl amino acid.

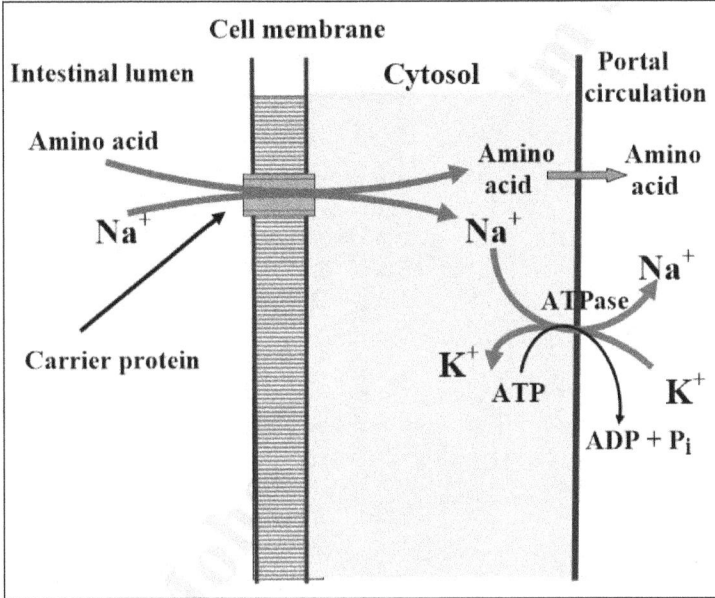

Figure 5.4: Carrier Protein Transport System.

☆ γ-glutamyl amino acid releases amino acid in the cytosol of intestinal mucosa cells with formation of 5- oxoproline that is used for regeneration of glutathione to begin another turn of the cycle.

Summary of Protein Digestion and Absorption

☆ Proteins taken in the diet are digested to amino acids in the stomach and small intestine.

☆ Gastric juice contains enzymes pepsin which digests protein in acid medium. It hydrolyses proteins to polypeptides.

Stomach=   **Dietary Proteins**
(Gastric Juice)

Pepsin digests in acid medium ↓

                                              Polypeptides

☆ In the small intestine, pancreatic and intestinal juices contain proteolytic enzymes. Pancreatic juice contains trypsin, chymotrypsin and carboxyl peptidase. They hydrolyse large protein molecule to smaller polypeptide.

Amino acids are absorbed into the portal vein and transported to the liver. From there they enter the general bloodstream.

Stomach- Protein is partially digested by the enzyme pepsin and hydrochloric acid.

Small Intestine- Final digestion of protein to amino acids.

Pancreas- Protein digestion by enzymes

Colon- Little dietary protein is present in feces.

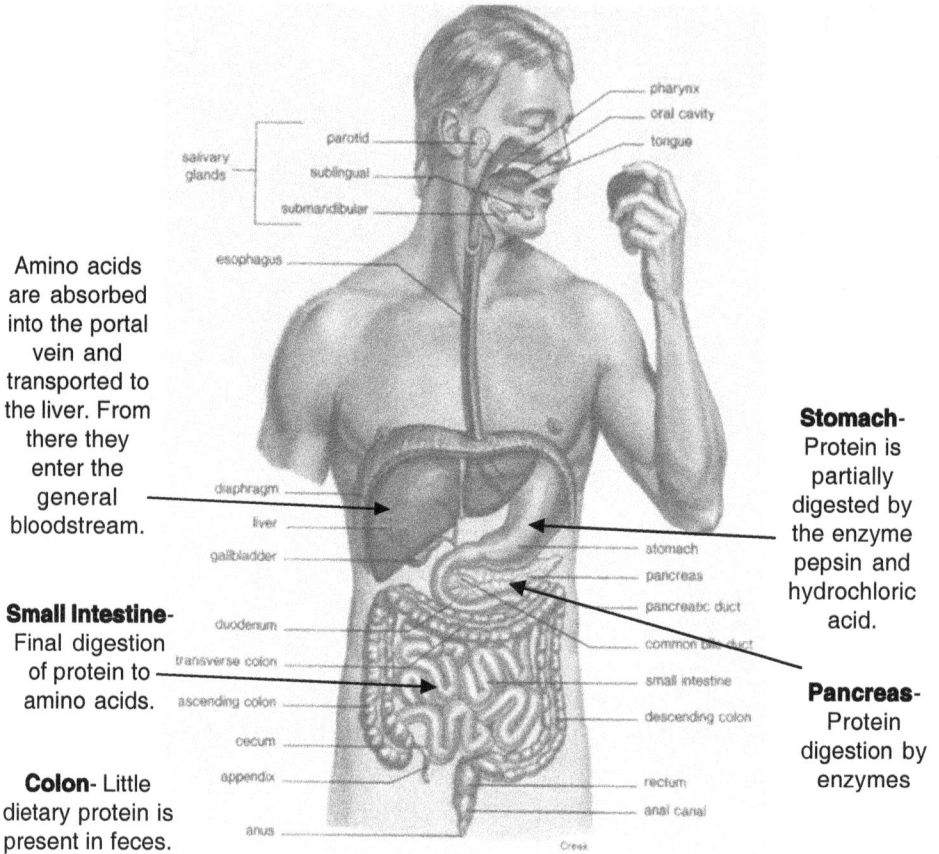

Figure 5.5: Summary of Protein Digestion and Absorption.

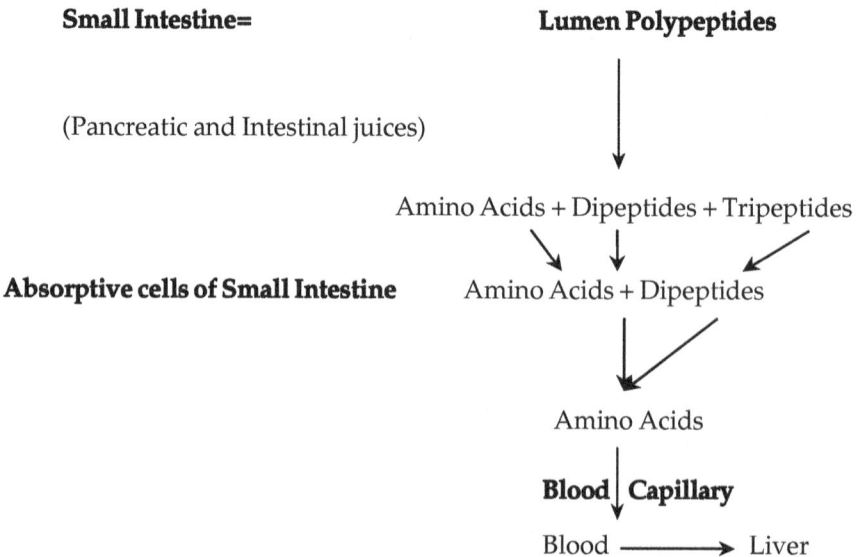

Small Intestine=

Lumen Polypeptides

(Pancreatic and Intestinal juices)

Amino Acids + Dipeptides + Tripeptides

Absorptive cells of Small Intestine

Amino Acids + Dipeptides

Amino Acids

Blood Capillary

Blood ⟶ Liver

5.7. Health Concerns Related to Protein Intake

Proteins play important roles in fluid balance, growth, immune function, and the transport of other nutrients. Thus, protein deficiency can severely compromise these functions.

Defect in Enzyme Synthesis

5.7.1. Oxoprolinuria

This is a disease caused by a defect in glutathione synthetase enzyme. It is characterized by accumulation of 5-oxoproline in blood and hence excreted in urine. It is usually associated with mental retardation.

5.7.2. Deficiency

Diet of many people living in developing or underdeveloped countries is insufficient in dietary protein so they may suffer with poor health. On the other end, the people of developed countries tends to eat more protein than required either from dietary sources or by consuming protein or amino acid supplements. Sufficient amounts of protein is required for maintaining good health, but too little or too much of it can have serious health effects on individuals. It can be called as **Protein-Energy Malnutrition (PEM).**

However, protein deficiency hardly develops as an isolated condition. Instead it most often occurs in combination with a deficiency of energy (and other nutrients), resulting in a condition known as protein-energy malnutrition (PEM), or protein-calorie malnutrition (PCM). PEM is a public health concern which can influence people of all ages in many developing or underdeveloped areas of the world where diets is low in protein and energy. The individuals with poor livelihood, and/or on restrictive diet problems, anorexia nervosa, or debilitating diseases (like cancer or AIDS), some hospitalized patients with protein malnutrition are also at higher risk of developing PEM because of poor prior health, low dietary intakes, and increased protein requirements for recuperation from operation, trauma, and/or disease. Malnourished patients encounter higher health complication and even death. Most devastating effects of PEM are observed in children because they have less protein intake. As inadequate protein and energy can hamper the normal growth of children, and many of them develops diarrhea, infections, and other diseases and because of which they can die early in life. PEM often develops as either marasmus or kwashiorkor. These conditions differ from each other.

5.7.2.1 Kwashiorkor

From birth, an infant in developing areas of the world is usually breastfed. Often by the time the child is 12 to 18 months old, the mother is pregnant or has already given birth again. The mother's diet is usually so marginal that she cannot produce sufficient milk to continue breast feeding the older child. This child's diet then abruptly changes from nutritious human milk to starchy roots and cereals etc. These foods have low protein densities, compared with their energy content. Additionally, the foods are usually high in plant fibers and bulk, makes it difficult for the child to consume enough to meet energy needs and nearly impossible to meet protein

Kwashiorkor	Marasmus
✰ Kwashiorkor is a word from Ghana that means "the disease that the first child gets when the new child comes."	✰ The word marasmus means "to waste away."
✰ Condition occurring primarily in young children who have an existing disease and consume a marginal amount of energy and severely insufficient protein.	✰ Condition that results from a severe deficit of energy and protein, which causes extreme loss of fat stores, muscle mass, and strength.
Characteristics of: ✰ Edema ✰ Mild to moderate weight loss ✰ Maintenance of some muscle and subcutaneous fat ✰ Growth impairment (60–80 per cent of normal weight for age) ✰ Rapid onset ✰ Fatty liver ✰ Apathy, diarrhea, lethargy, failure to grow and gain weight, infections, measles and withdrawal from the environment.	**Characteristics of:** ✰ Severe weight loss ✰ Wasting of muscle and body fat (skin and bones appearance) ✰ Severe growth impairment (less than 60 per cent of normal weight for age) ✰ Develops gradually ✰ Death

requirements. Many children in these areas also have infections and parasites that elevate protein and energy needs and leads to development of kwashiorkor.

The condition often reverses when the individual/patient gets diet sufficient in protein, energy, and other essential nutrients, the patient will begin to grow again normally when the infections of kwashiorkor are treated.

5.7.2.2 Marasmus

Over time, the severe lack of energy and protein results in a "skin and bones" appearance, with little or no subcutaneous fat. Marasmus usually develops in infants who either are not breastfed or have stopped breastfeeding in the early months. Often, the weaning formula used is incorrectly prepared because of unsafe water and because the parents cannot afford sufficient infant formula for the child's growing requirement. The latter problem may lead the parents to dilute the formula to provide more feedings, not realizing that this deprives the infant of essential calories, protein, and other nutrients.

When a mother is working or is away from home the child will be cared by others in that case, bottle-feeding is often necessary. So in this case, sufficient food with protein and other nutrients must be supplied. An infant with marasmus requires large amounts of energy and protein; unless the child receives them, full recovery from the disease may never occur. Most brain growth occurs between conception and the child's first birthday. If the diet does not support brain growth during the first months of life, the brain may not fully develop, resulting in poor cognitive and intellectual growth.

5.7.2.3 Marasmic Kwashiorkar

Marasmic kwashiorkar is marked protein deficiency and marked calorie insufficiency, sometimes it is referred to as the most severe form of malnutrition characterized by extreme weight loss, weakness, with features of kwashiorkar. This may also be secondary to other conditions such as chronic renal diseas or cancer cachexia in which protein energy wasting may occur.

5.7.3 Excess of Dietary Protein is it Good for Health?

In addition to recommending adequate protein consumption, the Food and Nutrition Board also suggests that protein intake not exceed 35 per cent of energy intake. Diets containing an excessive or disproportionate amount of protein do not provide additional health benefits. Instead, high protein intakes may increase health and disease risks.

1. Kidneys are responsible for excreting excess nitrogen as urea. Excesssive protein have a adverse effect on our kidneys. Thus, high-protein diets may overburden the kidney's capacity to excrete nitrogen wastes.

   Body requires sufficient amount of water to dilute and excrete urea, so inadequate fluid intake can increase the risk of dehydration as the kidneys use body water to dispose of the urea. So, this would have a serious impact for people with impaired kidney function. Therefore, lower-protein diet with adequate fluid intake is suggested for these individuals to help preserve kidney health.

2. When an individual takes excessive protein intake from animal proteins, then generally the diet will have less of plant-based foods. Consequently the diet will be low in fiber, some vitamins (vitamins C, E and folate), minerals (magnesium and potassium), and beneficial phytochemicals.

3. Animal proteins are often rich in saturated fat and cholesterol. As a result with high proteins from animal source will pose an increased the risk of cardiovascular disease.

4. High-protein diets also may increase urinary calcium loss and eventually leading to a loss of bone mass and an increased risk of osteoporosis. Chances of this to happen are less in individuals with adequate calcium intakes. These findings are controversial.

5. Our body obtain amino acids from dietary sources of proteins in proportions required for body functions and to prevents amino acid toxicity, especially for methionine, cysteine, and histidine—the most toxic amino acids. When individual amino acid supplements are taken, chemically similar amino acids will compete for absorption, resulting in amino acid imbalances and toxicity risk. This condition is normally observed in athletes, where health risks are associated with excess protein and amino acid supplementation.

5.8. How to Improve the Protein Quality?

5.8.1 Supplementation

supplementation of partially complete/incomplete proteins with complete proteins. Supplementation of limiting essential amino acid with protein concentrates high in amino acids can be done. For *e.g.* cereals are low in lysine. This also include replacement of the normal cereal grain with its high lysine mutant counterpart. Three high lysine cereals are not available corn, barley and sorghum. *e.g.* Kheer.

5.8.2 Complementation

The quality of dietary proteins depends on the pattern of essential amino acids it supplies. The best quality protein is the one that has an essential amino acid pattern very close to the pattern of the tissue proteins. Egg proteins and human milk proteins meets these guidelines and as classified as high quality proteins and serve as reference protein for defining the quality of proteins. The proteins of animal foods are (milk, meat, fish etc.) comparable with egg in their essential amino acid composition so, are considered as good quality proteins and have more digestibility. On the other hand, plant proteins are of poorer quality as compared for egg protein as EAA composition is not well balanced and EAA are not present in optimal level present in egg. For example, in comparison with egg protein, cereal proteins are poor in amino acid lysine and pulses and oilseed proteins are rich in lysine but are poor in sulphur containing amino acids. Individually such proteins are called as incomplete proteins. However, this deficiency of particular amino acid of any vegetable food can be conquered by judious blend with other vegetable foods which may have adequate level of that limiting amino acid. Thus amino acid composition of these proteins will complement each other and the resulting mix will have an amino acid pattern better than either of the constituent proteins of the mixture. This practice is usually followed to improve vegetable proteins quality.

Thus, a deficiency of an amino acid in one can be made-up to an adequate level in another, if both are consumed together. A protein of cereals deficient in lysine, content have a mutually supplementary effect, a combination of cereal and pulse in the ratio of 5:1 has been found to give an optimum combination. Ex. Balahar which is a combination of cereal and pulses, this is provided in aaganvadi and schools in mid day meal.

5.8.3 Fortification

This include addition of commercial amino acid to a product where the particular amino acid is absent.

5.8.4 Improvement of Protein Quality and Quantity by Gene Therapy.

Questions

Q. 1.   Which causes hydrochloric acid (HCl) release from gastric cells?

a. Gastrin   b. Secretin   c. Histamine   d. Acetylcholine

Q. 2. Which pancreatic enzyme is an exopeptidase?

a. Carboxypeptidase   b. Pepsin

c. Trypsin   d. Aminopeptidase

Q. 3. Which hormone stimulates bicarbonate secretion from the pancreas into the SI?

a. Motilin   b. Gastrin   c. Acetylcholine d. Secretin

Q. 4. Which is true of peptide absorption?

a. Occurs by facilitated diffusion

b. Requires sodium only

c. Tetrapeptides can be absorbed

d. Requires sodium and hydrogen

References

1. Wardlaw's, Perspectives in Nutrition, 8th Edition, McGraw-Hill Companies, ISBN 978–0–07–296999–3.

2. Shakuntala Manay, M. and Shadaksharaswamy, M., (1987). Foods-Facts and Principles, New Age International (P) Publishers Ltd., Chennai.

3. Nutrition by David C. Nieman, Diane E. Butterworth and Catherine N. Nieman, 1990 edition, Wm. C. Brown Publisher.

4. Bamji S. Mehtab *et al.,* Textbook of Human Nutrition, 3rd edition, 2010. Oxford and IBH Publishing Co. Pvt. Ltd., New Delhi. ISBN-978-81-204-1742-7.

5. Gopalan.C *et al.,* 2012, reprinted Nutritive Value of Indian Foods. NIN, Hyderabad.

6. 2010 revised Recommended Dietary Allowances suggested by Indian Council of Medical Research.

7. http://www.britannica.com/EBchecked/topic/479680/protein/72547/Classification-of-proteins

8. http://www.preservearticles.com/201105156692/notes-on-the-compositions-and-classification-of-proteins.html

9. http://www.preservearticles.com/201012251622/proteins-classification.html

10. http://www.google.co.in/imgres

11. http://www.google.co.in/#q=what+are+protein

12. http://ghr.nlm.nih.gov/handbook/howgeneswork/protein

13. http://en.m.wikipedia.org/wiki/proteinogenic_amino_acid

6
Lipids

6.0 Introduction

General Characteristics

Lipids that are solid at room temperature are called as fats (for example butter) and those that are liquid are called oils (for example vegetable oil). Lipids have following features:

☆ They are organic compounds made up of carbon, hydrogen and oxygen.

☆ Lipids as a class also have a lower ratio of oxygen to carbon compared with carbohydrate, proteins or alcohols.

☆ They rarely occur in on organism in free state, but are usually combined with proteins (lipoproteins) or with carbohydrate as glycolipids.

☆ They provide 9 kcal/g energy *i.e.* they supply more energy than carbohydrate, proteins.

☆ Lipids do not have repeating monomeric units *i.e.* they are not polymeric in nature.

Although lipids form a diverse group even then they share two main characteristics:

1. They dissolve in organic solvents such as chloroform, benzene and ether etc.

2. They are not readily soluble in water.

Fatty acids are long chain hydrocarbon containing a terminal carboxyl group and are found in most lipids in the body and in the lipids in food. Fatty acid is an organic acid composed of a chain of carbon atoms with attached hydrogen atoms; it has an acid group at one end and methyl group at the other end. These are normally found as esters in natural fats and oils, however they do occur in unesterified form when fatty acids bound to certain proteins.

Fatty acids with lesser number of carbon atoms (4 to 8) are present in milk fats, whereas those of intermediate chain length (10 to 14) and between 16 to 20 carbon atoms are found in most of the animal and vegetable fats. Most of the fatty acid especially those which occur in natural fats, contain an even number of carbon atoms, usually more than 14 carbon atoms, while most of the fatty acids have either 16, 18 or 20 carbon atoms, structure of fatty acid is shown below :

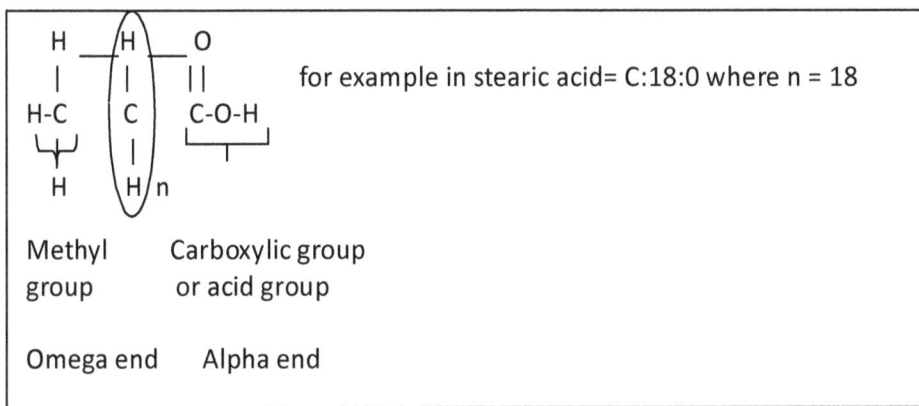

```
   H    /H\    O
   |    | |    ||        for example in stearic acid= C:18:0 where n = 18
  H-C   C    C-O-H
   |    ||     
   H    \H/n    

Methyl      Carboxylic group
group       or acid group

Omega end    Alpha end
```

Cis and Trans Fatty Acid

Monounsaturated and polyunsaturated fatty acids can exist in two isomeric forms namely cis and trans isomer, these isomers share the same chemical formula but have different chemical structures. Unsaturated fatty acids exhibit geometric isomerism.

The cis form has the hydrogen atoms on the same side of the carbon-carbon double bond. Plant origin foods contain cis fatty acids mainly. The trans form has the hydrogen atom on opposite side of the carbon – carbon double bond and are present in oil of deep fat frying, partially hydrogenated vegetable shortenings. Oleic acid contains cis form and elaidic acid contains trans form.

Presence of trans fatty acid in membranes make the membranes stiffer, this stiffness may reduce the function of cell membranes receptors that clears cholesterol from the blood stream this is the reason; why the trans fatty acids and saturated fatty acid raise serum cholesterol level in blood. Trans fatty acid content must be declared on the food label so that the consumer can make correct choice before purchasing that food item.

General Feature of Fatty Acid Structure

Carbon- carbon double bonds or unsaturation is found in naturally occurring fatty acids. There may be one double bond or many up to six in important fatty acids. Fatty acids with one double bond are the most prevalent in the human body, comprising about half of the total fatty acids. Fatty acids with two or more double bonds occur in lesser quantities, but are extremely important. When double bonds occur they are almost always cis, if there is more than one double bond then they occur at three carbon intervals forming divinyl methane pattern as given below:

$$-C=C-C-C=C-$$

Double bond pattern of polyunsaturated fatty acids in the divinyl methane pattern (showing the cis configuration) is depicted below:

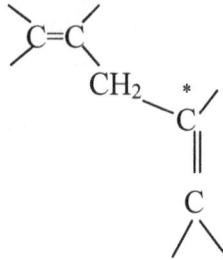

It is so named because it is as if a methane carbon (in the center) is attached to two vinyl group (carrying the double bonds). The pattern may be repeated to yield fatty acids with many double bonds. The above structure is the reason why omega fatty acids have double bond after 3, 6 or 9 carbons.

6.1 Lipid Classification

6.1.1 Lipids Classification on the Nature of Fatty Acid

1. Saturated Fat

If all bond between the carbon in the fatty acid structure are single bonds then a fatty acids is classified as saturated fatty acid and the fat containing such fatty acids as saturated fats. Saturated means that all the carbons are saturated with hydrogen's (a fatty acid with no carbon—carbon double bonds). Saturated fats are mostly solid at room temperature and become liquid when the temperature is increased (for example like ghee, butter etc.).

Animal fat, coconut oil, palm kernel oil, chocolate milk, cheese, butter, cream, beef, veal, palm oil, lard, pork, chicken are often source of saturated fatty acid and fat obtained from these sources is called as saturated fat.

Based on the carbon atoms saturated acids can be classified as below:

Saturated fatty acids are classified they have in their structure as summarized in table 6.1 below. Saturated fatty acids with more than 24 carbon atoms seldom occur in food triglycerides, but can occur in waxes. C_4-C_{10} occurs mainly in milk fat. These with carbon chain length from C_{12}-C_{24} are found in most animal and plant fats.

2. Unsaturated Fat

If bond between the carbon in the fatty acid structure are double bonds either singly or in multiple then a fatty acids is classified as unsaturated fatty acid and the fat containing such fatty acids as unsaturated fats. Unsaturated fats are mostly liquid at room temperature example include mustard oil, olive oil etc.

Table 6.1: Saturated Fatty Acid and their Sources

Fatty Acid	No. of Carbon Atoms	Source	State at Room Temperature
1.1. Short Chain: Fats containing C_4 to C_8 atoms			
Butyric acid	$C_4:0$	Butter	Soft or liquid
Caproic acid	$C_6:0$	Butter, coconut oil	Soft or liquid
Caprylic acid	$C_8:0$	Butter, coconut oil	Soft or liquid
1.2. Medium Chain: Fats containing C_{10} to C_{14} atoms			
Capric acid	$C_{10}:0$	Butter, coconut oil	Soft or liquid
Lauric acid	$C_{12}:0$	Butter, coconut oil	Soft or liquid
Myristic acid	$C_{14}:0$	Butter, vegetable foods	Soft or liquid
1.3. Long Chain: Fats containing C_{16} and more carbon atoms			
Palmitic acid	$C_{16}:0$	Most vegetable and animal fats	Solid
Stearic acid	$C_{18}:0$	Most vegetable and animal fats	Solid
Arachidic acid	$C_{20}:0$	Butter, lard, peanut oil	Solid
Behenic acid	$C_{22}:0$	Vegetable oils	Liquid

Based on number of double bonds in the fatty acids structure, unsaturated fatty acids are classified as:

2.1 Monounsaturated Fatty Acids (MUFA)

If the fatty acid has one carbon-carbon double bond for example as in oleic acid, $C_{18}:1$, ω-9 (here 18:1 means that the carbon chain is of 18 carbon having one double bond at ω-9 position) as found in olive oil (77 per cent MUFA) and canola oil (62 per cent MUFA). Structure of oleic acid is shown below:

```
    H   H   H   H   H   H   H   H           H   H   H   H   H   H   H   O
    |   |   |   |   |   |   |   |           |   |   |   |   |   |   |   ||
H—C—C—C—C—C—C—C—C—C = C—C—C—C—C—C—C—C—COH
    |   |   |   |   |   |   |   |   |   |   |   |   |   |   |   |   |   |
    H   H   H   H   H   H   H   H   H   H   H   H   H   H   H   H   H
```

Other monounsaturated acids are palmitoleic acid ($C_{16}:1$) found in olive oil, fish oil and beef fat, erucic acid ($C_{22}:1$) found in rape seed oil and canola oil.

MUFA is also found in avocados, almonds, margarine, peanut oil, peanuts (49 per cent MUFA), cottonseed oil, had dock, safflower or kardy oil (13 per cent MUFA).

```
    H   H   H   H   H           H           H   H   H   H   H   H   H   O
    |   |   |   |   |           |           |   |   |   |   |   |   |   ||
H—C—C—C—C—C—C=C—C—C = C—C—C—C—C—C—C—C—COH
    |   |   |   |   |   |   |   |   |   |   |   |   |   |   |   |   |   |
    H   H   H   H   H   H   H   H   H   H   H   H   H   H   H   H   H
```

2.2 Polyunsaturated Fatty Acids (PUFA)

If two or more bonds between the carbons are double bonds, the fatty acids is called as PUFA this includes linoleic acid, C_{18}:2, ω-6 and α-linolenic acid, C_{18}:3, ω-3. PUFA is found in soft margarine, sesame oil, mayonnaise, soybean oil, corn oil, sunflower oil, safflower oil, fish fat. Structure of linoleic acid is shown below:

Fatty acid composition of some edible oils and fats is given in table 6.2. PUFA can further be divided as omega-3, 6 and 9 fatty acid (ω-3, ω-6 and ω-9) based on the position of first double bond.

1. **Omega 3 (ω-3) fatty acids:** unsaturated fatty acid with the first double bond located on the third carbon atom from the methyl end (-CH_3) of the molecule, for example α-linolenic acid (C_{18}:3, ω-3) as found in soyabean oil, eicosapentaenoic acid (C_{20}:5, ω-3) found in fish oil and docosahexaenoic acid (C_{22}:6, ω-3) as found in shell fish.

2. **Omega 6 (ω-6) fatty acids:** unsaturated fatty acids with the first double bond located on the sixth carbon atom from the methyl end (-CH_3) of the molecule, for linoleic acid (C_{18}:2, ω-6) found in safflower, corn, soyabean and cotton seed oil, arachidonic acid (C_{20}:4, ω-6) as found in fat and phosphate fractions of animal tissues of liver, lard and meat.

3. **Omega 9 (ω-9) fatty acids:** unsaturated fatty acid which has the first double bond located on ninth carbon atom from the methyl end of the fatty acid for example as found in oleic acid.

6.1.2 Three Major Classes of Lipids Namely Triglycerides, Phospholipids and Sterols

1. Triglycerides (TGL)

TGL are the acyl derivatives of glycerol and are the most common type of lipid found is the body and in foods. Each triglycerides molecule consists of a glycerol with three fatty acids attached to it by ester bonds as shown below:

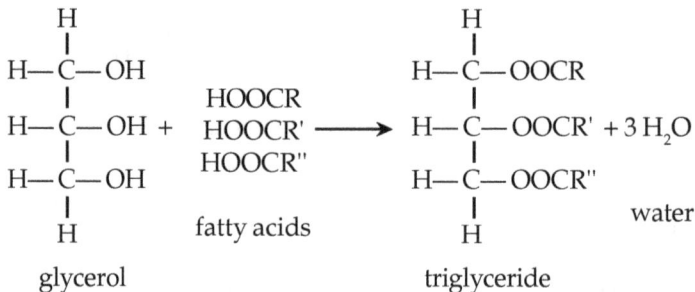

glycerol   fatty acids   triglyceride   water

About 95 per cent of all fatty acids, both in food and body all in the form triglycerides. Thus, butter, body fat, soyabean oil all contains primarily triglycerides.

2. Phospholipids (PL) or Phosphoglycerides

Phospholipids play important role in cell membranes and considerably influence its structure and function. These are also involved in the transport of other lipids in

Table 6.2: Fatty Acid Composition of some Edible Oils and Fat

Fat/Oil	Palmitic 16:0	Stearic 18:0	Arachidic 20:0	Behenic 22:0	Lignoceric 24:0	Total Saturates	Palmitoleic 16:1	Oleic 18:1	Total Mono-saturates	Linoleic 18:2	α-Linolenic 18:3	Total Poly Unsaturates
Coconut oil	7.8	2.3	–	–	–	89.5*	–	7.8	7.8	2.0	–	2.0
Corn oil	10.7	1.7	–	–	–	12.7	–	29.6	29.6	57.4	–	57.4
Groundnut oil	12.6	1.7	4.2	2.1	0.3	20.9	1.4	47.9	49.3	29.9	–	29.9
Mustard oil	2.9	0.9	6.9	–	–	10.7	0.6	8.9	56.0**	18.1	14.5	32.6
Olive oil	12.5	2.3	–	–	–	14.8	–	74.5	74.5	10.0	–	10.0
Palm oil	42.0	4.3	–	–	–	46.3	–	43.7	43.7	10.0	–	10.0
Palmolein	42.3	3.5	–	–	–	47.7	0.4	41.0	41.4	10.3	0.3	10.6
Rice bran oil	19.5	2.6	–	–	–	22.1	–	41.0	41.0	34.3	1.4	35.7
Safflower oil	7.8	2.1	0.8	–	–	10.7	–	16.7	16.7	73.5	–	73.5
Gingelly oil	9.7	4.0	–	–	–	13.7	0.1	41.2	41.3	44.5	–	44.5
Soyabean oil	9.8	2.4	0.9	–	–	13.1	–	28.9	28.9	50.7	6.5	57.2
Sunflower oil	5.6	2.2	0.9	–	0.4	9.1	–	25.1	25.1	66.2	–	66.2
Butter	29.4	13.5	–	–	–	69.4*	–	28.0	28.0	2.5	–	2.5
Lard	26.2	20.0	–	–	–	46.2*	–	45.2	45.2	11.0	–	11.0
Tallow	27.9	21.3	–	–	–	54.9*	–	40.9	40.9	4.2	–	4.2

*Includes lower chain fatty acid

** Includes 46.5 per cent of erucic acid (22:1).

Source: Nutritive value of Indian foods, C. Gopalan *et al.,* NIN, ICMR.

the blood stream. These are usually complexed with proteins as lipoproteins in their natural state. Humans do not have to eat phospholipids as such because our body can make these when and where it is necessary. Phospholipid intake is not associated to any major disease state.

PL are different from triglycerides in that one of the terminal hydroxyls of glycerol is esterified with phosphoric acid and the other two hydroxyls by fatty acids. The fatty acid occupying n-1 and n-2 positions vary in chain length and in degree of unsaturation. Acids in position 2 are usually highly unsaturated and in position 1 can be saturated or monoenoic. The phosphoric acid in position 3 is in turn, esterified to another moiety "X" in the structure below:

$$
\begin{array}{c}
O \\
| \\
O \quad\quad H_2C-O-O-R_2 \\
\| \quad\quad\quad | \\
R_1-C-O-C-H \quad O \\
| \quad\quad\quad \| \\
H_2C-O-P-O-X \\
| \\
O^-
\end{array}
$$

General structure of phospholipid

The "X" may be choline, ethanolamine, serine or inositol. Phosphoglycerides containing these components are all found in foods. The phospholipids containing choline is known as lecithin. Lecithin is an important component of all cell membranes and participates in fat digestion in the small intestine. Egg yolk is rich in lecithin and is thus use as emulsifier. Other sources are liver, wheat germ and peanuts.

3. Sterols

These are another type of lipids, which consist of a multi ring (steroid) structure and at least one hydroxyl group (-OH). Sterols are unlike other lipids in their structure but qualify to be lipid because of their physical properties. Sterols are unlike other lipids in their structure but differ because of their physical properties. The most common sterol are cholesterol, ergosterol, vitamin D, estrogen and other hormones are found only in animal derived food constituents we eat.

Cholesterol is an essential structural component of cell membrane and the outer layer of particles that transport lipids in the blood. Cholesterol is also precursor of bile acids, which are needed for fat digestion. It is also a part of important hormones such as corticosteroids, estrogens, testosterone, calcitriol. Neither cholesterol has a unique role in food nor is an essential part of an adult's diet. Cholesterol is found in foods of animal origin which contain cholesterol naturally such as egg, meat, fish, poultry and dairy products. Foods of plant origin do not contain cholesterol, they do make other sterols, such as ergosterol (a form of vitamin D) and sitostanol. Most people get about one-third of their cholesterol from the foods we eat and the rest is manufactured by our bodies.

6.1.3 Lipid Classification Based on Structure

Based on the structure lipids are classified into three main classes:

1. Simple Lipids

Chemically these are esters of fatty acids with alcohols. Oil and fats are common examples of this class of lipids which are esters of long chain fatty acids with glycerol. These esters called triglycerides or triacylglycerols. They are solid or semisolid at room temperature and are known as fats. These occur predominantly in animal fat and are generally triglycerides containing saturated fatty acids. On the other hand, triglycerides containing unsaturated fatty acids are liquid at room temperature and are known as oils. These are present in plant oil. Fish oil is an exception as it contains triglycerides which are esters of unsaturated fatty acids with glycerol hence it is called as oil.

Waxes are also simple lipids and are esters of fatty acids with long chain monohydric alcohols. These are widely distributed in nature. For example the leaves and fruits of many plants have waxy coatings, which protect them from dehydration and small predators. Feathers of birds and fur of few animals have similar coating which serves as a water repellent layer. Waxes are usually inert due to the saturated nature of the hydrocarbon chain. Carnauba wax which is obtained from carnauba palm is an example of a tough and water resistant wax. It contains esterified fatty dialcohols (diols about 20 per cent), hydroxylated fatty acids (about 6 per cent) and cinnamic acid (about 10 per cent). Rice bran obtained from milling of rice contains wax mixed with triglycerides.

2. Complex Lipids

Complex lipids are also esters of fatty acids with alcohol. However in addition to these two components they contain other groups, such as a phosphate group or a carbohydrate moiety. Depending upon the moiety attached, these are of the following types:

a) **Phospholipids** : These complex lipids contain a phosphate group, in addition to fatty acids and an alcohol. They frequently have nitrogen containing bases and other substituent as well. Phospholipids that contain glycerol as the alcohol are known as glycerol phospholipids and those having sphingosine as the alcohol moiety are called the sphingo phospholipids.

   In daily products, phosphotidyl ethanolamine (PE), phospahatidylinositol (PI), phosphatidylserine (PS), and phosphatidylcholine (PC) are important phospholipids. Whereas splingomyelin (SM), glucosylceramide (GLUCER), lactosylceronide (LACCER) are important dairy splingolipids.

   IGNOU

b) **Glycolipids**: These lipids contain a carbohydrate group in addition to fatty acids and an alcohol. The sugar group in glycolipids is usually galactose though sometimes these may contain glycose. The alcohol part is either sphingosine or glycerol.

   IGNOU

c) **Other complex lipids:** This category consists of lipids such as sulpholipids, aminolipids, lipoproteins containing sulphur, amino and protein molecule respectively.

3. Derived Lipids

These are obtained from the complex lipids by hydrolysis and include fatty acids (saturated as well as unsaturated), glycerol, steroids, lipids soluble vitamins like vitamin A, D, E, K and prostaglandins.

6.1.4 Lipids Classification Based on the Ability of Body to Synthesize Fatty Acid-Essential Fatty Acid (EFA) and Non Essential Fatty Acid

Essential Fatty Acid

Cells in the human body can produce carbon – carbon double bonds in a fatty acid only after the ninth carbon from the methyl end. This means that human cell doesn't have the capacity to place double bonds between the ninth carbon and the methyl end. Because of this reason and essentiality in one's diet, the omega-3 and omega-6 PUFA (α-linolenic acid and linoleic acid respectively) are called essential fatty acids. Functions and food sources of EFA are discussed in Table 6.3 below:

Table 6.3: Essential Fatty Acids and their Sources

Essential Fatty Acid	Function	Food Sources	State at Room Temperature
Omega 3: alpha linolenic acid	It decrease inflammation responses, blood clotting and plasma triglycerides level	Cold-water fish (like salmon, tuna, sardines, mackerel), walnuts, flaxseed, canola oil, soybean oil	Liquid
Omega 6: Linoleic acid	It regulates blood pressure and increases blood clotting	Beef, poultry, safflower oil, sunflower oil, cornoil	Solid to liquid

Because of this reason omega 3 and omega 6 fatty acids are to be obtained by ingesting them from diet for maintaining good health. These fatty acids form important part of vital body structures, perform important role in immune system function and vision, help form cell membranes, and produce hormone like compounds called as eicosanoids (these are synthesized from PUFA such as arachidonic acid, this include prostaglandins, thromboxanes, leukotrienes). Eicosanoids made from omega-3 fatty acids tend to decrease blood clotting, blood pressure and inflammatory responses in the body. Other important roles of eicosanoids in the body :

☆ Involved in maintaining normal kidney function, fluid balance and transporting oxygen from RBC's to body tissues.

☆ Regulate rate of cell division, which may help prevent certain cancers or slow the growth of existing tumors and help prevent cancer from spreading to other parts of the body.

☆ Direct hormones to their target cells.

☆ Regulate the flow of substances into and out of the cells.

☆ Regulating ovulation, body temperature, immune system function, and hormone synthesis.

Dietary guidelines for intake of omega fatty acids are established in Canada only. In 1989, published data on RDA states that adult need to consume about 1 per cent to 2 per cent of their total energy intake from linoleic acid and upto 10 per cent of energy intake from a combination of linoleic acid and α-linolenic acid is a safe upper amount. Plant seed oils like sunflower, safflower, cotton seed, soybeans, corn oil contain over 50 per cent of their fatty acid as PUFA. Plant oils should provide about 4 per cent of our total energy intake. Salad dressing, mayonnaise are common ways for incorporating essential fatty acids in our diet. Different omega acids are discussed below:

1. **Eicosapentanenoic acid (EPA):** the fatty acids eicosapentaenoic acid (EPA) and, subsequently, docosahexaenoic acid (DhA) are made from alpha-linolenic acid. EPA is an omega-3 fatty acid with 20 carbons and five carbon carbon double bonds. It is present in large amount in fish oil and is also synthesized in the body from alpha-linolenic acid.

2. **Docosahexaenoic acid (DHA):** An omega-3 fatty acid with 22 carbons and 6 carbon – carbon double bonds. It is present in large amount in fish oils and is also synthesized in the body from alpha-linolenic acid.

3. **Arachidonic acid**: An omega-6 fatty acid with 20 carbons and four carbon-carbon double bonds.

4. **Prostaglandins:** A group of eicosanoids that produce diverse hormone like effect in the body.

5. **Thromboxane:** A stimulant of blood clotting made in certain blood cell, is an eicosanoid.

6.1.5 Fatty Acid Classification Based on Nutritional Value

1. **Non Essential fatty acid:** They are generally MUFA (are found in olive oil, canola oil, peanut oil and are liquid at room temperature).

2. **Essential fatty acid:** They are generally PUFA (are found in sunflower oil, corn oil, safflower oil, fish oil and are liquid at room temperature).

3. **Semi Essential/Conditionally/Partially Essential fatty acid:** If precursor for their formation is not present then they become essential fatty acid. One example is that our body can use linoleic acid to synthesize arachidonic acid which helps to maintain the structure and function of all membrane.

6.1.6 Fat Classification on the Basis of Visibility

Food contains many components having diverse structures and functions. Food Lipids are consumed as we eat food, food may contain lipid in the form of visible and invisible fats.

1. **Visible fats:** are called so because these are the fat component that can be seen by human after their separation from the original plant or animal sources this include vegetable oils, butter etc. These are visible as their fat concentration is more than 80 per cent.

2. **Invisible fats:** are constituents of basic foods as present in fruits,vegetables, pulses, cereal, milk, cheese, meat, etc. and are consumed without being separated from food. This contributes to nearly 60 per cent of dietary fat.

6.2 Dietary Sources

Most cereals, vegetables, fruits contain very little fat. However some fruits and vegetables like corn, avocado and the fruit palm are also rich sources of fats/oils.For example maize contains 3.6 per cent, wheat germ(7.4 per cent), almond(58.9 per cent),buffalo milk (8.8 per cent), avocado(22.4 per cent),and mutton(13.3 per cent) fat/oil. The largest sources of vegetable oils are seeds of annual plants such as soybean (18-20 per cent), cotton seed (18-20 per cent), peanut (45-50 per cent), linseed or flax (~40 per cent), sunflower (35-45 per cent), safflower (30-35 per cent), mustard (28.76g/100g), coconut (65-68 per cent). Few sources of dietary fat/oil are listed below in table 6.4.

Table 6.4: Fat and Fatty Acid Composition of Common Foods (g/100g)

Food Source	Total Fat	Total Saturates	Total Monounsaturates	Total Polyunsaturates
Maize	4.8	0.90	1.1	2.25
Bajra	5.5	1.24	1.2	2.33
Bengal gram	6.9	0.57	1.2	3.7
Soyabean	20.0	2.2	5.4	11.8
Coconut	40.0	36.0 (includes lower chain fatty acid)	3.2	0.6
Groundnut	40.0	8.8	21.0	10.2
Mustard	40.0	2.0	27.0 (includes erucic acid 22:1)	8.5

Source: Nutritive value of Indian foods, C. Gopalan *et al.*, NIN, ICMR.

Sources can also be classified as a source of saturated or unsaturated fat as discussed below:

1. **Sources of saturated fat:** All butter, coconut oil etc. they contain generally short or medium chain fatty acid

2. **Sources of unsaturated fat:**

    2.1 **Monounsaturated Fatty Acid (MUFA):** olive oil (77 per cent MUFA), canola oil (62 per cent MUFA) are good sources of MUFA. MUFA is also found in avocados, almonds, margarine, peanut oil, peanuts (49 per cent MUFA), cottonseed oil, had dock, safflower or kardy oil (13 per cent MUFA).

2.2 Polyunsaturated Fatty Acid (PUFA): Plant seed oils like sunflower, safflower, cotton seed, soybeans, corn oil contain over 50 per cent of their fatty acid as PUFA. PUFA is found in soft margarine, sesame oil, mayonnaise, soybean oil, corn oil, sunflower oil, safflower oil, fish fat. PUFA in oils like groundnut oils, safflower or sunflower oil have been shown to prevent an increase in serum cholesterol on a high fat diet and are thus considered antiatherogenic.

6.3 RDA for Fat

As such there is no RDA for dietary fat. But to obtain the essential fatty acid adults should consume about 4 per cent of energy from plant oils incorporated into foods and eat fish about twice a week. As per recommendations of American Health Association (AHA):

☆ Total fat intake should be less than 30 per cent of energy intake.

☆ Saturated fat intake should be less than 10 per cent energy intake.

☆ Polyunsaturated fat intake should be sufficient to maintain the individual.

☆ ICMR has given a ratio for MUFA: PUFA: SATURATED FAT=1:1:1. Out of the total fat intake of 30 per cent, 10 - 20 per cent should be from invisible fat sources and less than 20 per cent should be from visible fat source.

☆ The Upper limit of fat in the diet should not exceed 30 per cent of calories *i.e.* it should be less than 80g/day.

☆ The fat requirement is summarized as below:

Group	Fat Adequate for Meeting Daily Requirement
For a normal person (Children, Adolescence)	15-25 g/day
Pregnant Women	30 g/day
Lactating Mother	45 g/day

6.4 Functions of Lipids

1. Function of Triglycerides (TGL) in the Human Body

a) Triglycerides are the main chemical form of stored energy in the body. The ability to store triglycerides is essentially unlimited as adipose cells can increase about 50times in weight

    1.1 TGL is the storage form of energy

    1.2 Energy Density: 1g of fat provides 9 kcal/g energy.

    1.3 FFA released from triglycerides makes up the main fuel for muscles at rest and during light activity.

b) **Protecting the body**: Adipose tissue layer under the skin is composed mostly of triglycerides which protect the body especially breast, kidneys and also protects vital body organs from damage or bruises.

c) **Absorption of vitamins**: It aids in intestinal absorption of fat soluble vitamins – A, D, E, K. TGL and other lipids in foods carry fat soluble vitamin to the small intestine and in doing so it aids in absorption of these nutrients.

d) **Insulation**: It help maintain the inner temperature of the body, thus avoiding rapid changes in body temperature. About half of the adipose tissue acts as insulation. It is stored directly under the skin in the subcutaneous fat layer and is useful in controlling temperature and in creating pleasant appearance but excess of subcutaneous fat may not be appealing or healthy.

e) **Satiety and flavour:** As the dietary fat causes the stomach to empty slowly than either carbohydrates or protein, so it impacts satiety and help in feeling satisfaction after eating.

2. Functions of Phospholipids

a) Lecithin and other phospholipids are necessary components of body tissues but we don't have to consume phospholipids as such because our body can make them as and when necessary.

b) Phospholipids and lecithin are important component of all cell membrane and participate in fat digestion in the small intestine.

c) Most phospholipids functions as emulsifier.

d) Phospholipids like sphingomyelines and cerebrosides are found in brain and other nerve tissues thus making them important.

e) Phospholipids are also essential components of certain enzymes (Lipoprotein lipase and others).

f) Phospholipids are involved in lipid transport in the stomach, in intestine, in bloodstream and lymphatic system. The blood is the main medium for moving materials around in the body and is water based substance. Since lipids and water don't mix so the body must have some ingenious mechanism for transporting lipids in the bloodstream. **Lipoprotein is** a particle found in the blood containing which has a core of lipids with a shell of proteins, phospholipids and cholesterol. The outer shell around lipoproteins allows the lipid it is carrying to float freely in the water based blood, thus allows the transportation of lipids.

3. Function of Sterol Component

They are precursor of certain vitamins and hormones for example:

a) Cholesterol is a precursor for numerous hormones and bile acids. Cholesterol is also the precursor of vitamin D.

b) Lipids like phospholipids and cholesterol are involved in structured components and permeability of membranes in cells.

4. Protein Sparing Effect

Fats also spares body protein for their utilization for producing energy since sufficient amount of fat furnish the energy needed to carry body work.

5. Function of Lipids in Food Processing

a) Mono-, di-glycerides, phospholipids and a wide range of synthetic compounds are used as emulsifiers. Therefore the texture of foods can be modified by using mixture of fat blends and appropriate emulsifiers.

b) Lipids themselves or their breakdown products contribute to the flavor and aroma of food for example flavor of butter, milk and cheese is due to short chain fatty acids.

c) Many flavors associated with vegetables such as beans and cucumber are formed by the oxidative enzymes breakdown of unsaturated tissue Lipids.

6.5 Conditions Resulting due to Excess Dietary Fat

1. High proportion of saturated fats in the diet has been shown to cause considerable elevation of blood cholesterol when consumed in large amounts. Eating excess of dietary fat increases the chances of developing coronary heart diseases, cardio vascular diseases (CVD) and obesity.

   a) Cardiovascular diseases include heart attack, strokes, high blood pressure and atherosclerosis. Major risk factor for these is high blood cholesterol levels. CVD can result by consuming a high LDL cholesterol and Low HDL cholesterol diet. The consumption of total fat, saturated fat and dietary cholesterol are associated with elevations in blood cholesterol. A diet high in cholesterol, saturated fat, trans fatty acid can result in development of CVD.

   b) Atherosclerosis: Excess cholesterol in the blood gradually leads to its being deposited under the blood vessels lining resulting in a condition known as atherosclerosis in which the blood vessels are narrow and hardened. The coronary arteries supplying blood to the heart are affected and will result in coronary heart disease. This is a slow, progressive disease that begins in childhood and takes decades to advance. A high level of homocysteine (amino acid) may promote atherosclerosis.

   c) Obesity: Eating a fat rich diet or excessive dietary fat can result in the deposition of fat. As been observed today, much of the obesity is due to calorie rich foods that are high in fat. Research also shows that it costs the body less energy to store dietary fat as body fat (3 per cent of ingested calories) than to store carbohydrate as body fat (23 per cent of ingested calories). Calories from other source can also create obesity, but many researchers have suggested that a high carbohydrate low fat diet will help individuals to remain lean or lose excess fat.

2. Other harmful effects of fats/fatty acids:

   a) Cancer: A diet high on total fat may be a factor in the development of breast, colon, rectal, uterine, prostrate cancer.

   b) Diabetes: Individuals with diabetes have twice the risk of developing cardiovascular disease than those without the disease therefore a diet low in total fat, cholesterol and saturated fat is consistent with the guidelines of American Heart Association.

c) Some cyclic fatty acids inhibit the destruction of stearic acid to oleic acid. This can alter membrane permeability and leads to some diseases.

d) Abnormal lipid metabolism is the causative agent in the development of coronary dietary disease.

e) Some per oxidized products (produced during oxidative rancidity) of polyunsaturated acids could be toxic for human health.

3. **Hypertriglyceridemia** is a condition of high (hyper-) blood levels (-emia) of triglycerides which is the most abundant fatty molecule in most organisms. Hypertriglyceridemia itself is usually symptomless but elevated levels of triglycerides are related with atherosclerosis, increase the risk of acute pancreatitis, skin lesions known as xanthomas. There is a hereditary predisposition to both primary and secondary hypertriglyceridemia. Weight loss and dietary modification may be effective in treating hypertriglyceridemia however medication may be necessary to treat high levels of hypertriglyceridemia and hypertriglyceridemia coupled with the presence of other risk factors for cardiovascular disease.

4. **Hypercholesterolemia/hypercholesterolaemia** is a condition of the presence of high levels of cholesterol in the blood. It is a form of "hyperlipidemia" (elevated levels of lipids in the blood) and "hyperlipoproteinemia" (elevated levels of lipoproteins in the blood).

Cholesterol is a sterol and is insoluble in water, it is transported in the blood plasma within protein particles (lipoproteins). All the lipoproteins (HDL, LDL, VLDL) carry cholesterol. Higher levels of HDL cholesterol are protective but elevated levels of the lipoproteins other than HDL (termed as non-HDL cholesterol particularly LDL-cholesterol) are associated with an increased risk of atherosclerosis and coronary heart disease. Elevated levels of non-HDL cholesterol and LDL in the blood may be a effect of diet, obesity, inherited (genetic) diseases (such as LDL receptor mutations in familial hypercholesterolemia), or the presence of other diseases such as diabetes and an underactive thyroid.

Reducing dietary fat is recommended to reduce total blood cholesterol and LDL in adults. In people with very high cholesterol (*e.g.* familial hypercholesterolemia), as diet is often insufficient to achieve the desired lowering of LDL so, medications which reduce cholesterol production or absorption are usually recommended by doctors.

Deficiency Conditions

As a healthy practice one should eat a diet low in fat which help one to control body weight and hence control the chances of developing some of the harmful diseases like obesity, CVD etc. However, at the same time, an individual should also ensure that deficiency of fat or fatty acids doesn't develop in the one's body. Deficiency situation of lipids/fatty acids in the body can result either due to improper dietary fat or due to conditions resulting because of improper lipid metabolism in the body as discussed below:

1. If the small intestine is diseased, dietary fat may not be properly digested and absorbed. Under such situation, fat passes through the small intestine and enters the large intestine. The fat-soluble vitamin bound to unabsorbed fat also are carried into the large intestine by passing their absorption sites in the small intestine.

2. People with cystic fibrosis poorly absorbs fat and are thus at risk for developing deficiency of fat-soluble vitamin especially vitamin K.

3. Deficiency of essential fatty acids:

   If an individual follows, a diet following food pyramid for planning meal then that meal will provide enough essential fatty acids required by human body. As we just require only a table spoon of oil rich in polyunsaturated fatty acid to meet daily requirement of essential fatty acid, but if this amount is not consumed then deficiency symptoms of essential fatty acids can develop.

   In infant deficiency symptoms includes skin becoming flaky and itchy, sours may develop on the scalp, diarrhea and other symptoms such as infections are often seen, wound healing &retarded growth can also be seen. In adults, skin disorders and anaemia is observed majorly. EFA play a role in several metabolic reactions, deficiency of these fatty acid lead to a skin condition known as phrynoderma (toad skin) in which skin becomes rough and thick horny papules of the size of a pin head erupt in certain areas of the body, notably thighs, buttocks, arms and trunk. But some studies have shown that this condition responds to vitamin E and B complex more effectively than to EFA deficiency.

4. **Insufficient consumption:** Insufficient consumption of lipid (*i.e.* less than three percent of the total caloric intake) most often occur in infants who are fed on low fat or non fatmilk or formulas made from those, also parents stop there young kids to prevent obesity or over weight are some of the reasons for deficiency of lipids in the body.

   Nevertheless, parents need to be cautioned that limiting lipids is not meant for individuals under the age of 2 years, because essential fatty acids are needed in the diet particularly for growth stages. If child doesn't get sufficient EFA types his or her growth will be impaired.

5. If the diet lacks the lipids then deficiency of fat soluble vitamins (A, D, E, K) can occur in the individuals.

6. DHA and EPA (omega 3 fatty acid) deficiency can lead to neurological and vision abnormalities.

7. **Hypocholesterolemia** is the presence of abnormally low (hypo-) levels of cholesterol in the blood (-emia). According to the American Heart Association in 1994, only total cholesterol levels below 160 mg/dL or 4.1 mmol/l are to be classified as "hypocholesterolemia". However, this is not agreed on universally and some put the level lower. Although the presence hypercholesterolemia is related with CVD but it is not clear if a lower than

average cholesterol level is directly harmful. This condition is often observed in particular illnesses.

6.6 Lipid Digestion and Transport

6.6.1 Lipid Digestion or Transport in Body

There is wide variety of lipids to be transported into the body like triglycerides, cholesterol, steroids etc. All of these have to be transported in the body for their digestion for subsequent absorption, the process occur as below:

1. The stomach produces gastric lipase (an enzyme that digest fat). This enzyme acts primarily on triglycerides containing short and medium chain fatty acids such as those found in butter fat.

2. The digestion occurs only in stomach because gastric lipase requires an acidic environment to functions.

3. Pyloric sphincter opens up to transfer food from stomach to small intestine.

4. The action of gastric lipase is usually dwarfed by the action of pancreatic lipase in small intestine. Digestion of triglycerides containing long chain fatty acids such as those found in common vegetable oils, occurs in small intestine.

5. The hormone CCK (cholecystokinin, whose origin is in duodenum and jejunum part of small intestine) which release enzyme for protein digestion, simultaneously releases lipase from the pancreas for fat digestion.

   The hormone CCK cause contraction of gall bladder and flow of bile to duodenum, it causes secretion of enzymes rich in pancreatic fluid and slows gastric emptying.

6. In the small intestine this pancreatic lipase digests the triglycerides in smaller products, namely monoglycerides and fatty acids.

7. Pancreatic lipase enters the small intestine through the common bile duct in a concentration many times greater than desired. This makes fat digestion very rapid and thorough in right circumstances (which include the release of co-lipase from pancreas to aid in lipase enzyme action).

8. Bile which is made in the liver, stored in the gall bladder and released into the small intestine in response to CCK. Bile helps to emulsify the fatty substances in the small intestine, this increases the surface area which is exposed to digestive enzyme.

Bile is composed of cholesterol, lecithin and bile salts sterols. Large triglycerides are emulsified or broken down to smaller component/droplets by bile. Bile also organizes the digestive products of lipase action into micelles. This improves fat absorption because the digestive products of fat are transported mostly as micelles to the intestinal wall to begin fat uptake.

Triglycerides, glycerol, fatty acid and monoglycerides are incorporated with bile salts in a package called a micelle. Micelle delivers the contents of the intestinal cell membrane for transport into the body.

Dietary cholesterol is usually attached to a fatty acid in foods. This is first removed in the small intestine by lipase then cholesterol can also be incorporated into micelles. Again the micelle delivers the cholesterol to the intestinal cell for transport into the body. If some triglycerides are left undigested by chance then they will be trapped in dietary fibre and will pass from small intestine to large intestine for waste excretion.

6.6.2 Fat Absorption

1. Most of the products of fat digestion are fatty acids and monoglycerides. These are passively absorbed into the absorptive cells.

2. The carbon chain length of fatty acids and monoglycerides affects the manner of their absorption. If a fatty acid is a short or medium chain variety (*i.e.* less than 12 carbon atoms) it is water soluble and probably travels through the portal vein to the liver. If the fatty acid is a long chain variety (12 or more carbon atoms), it is re- formed into a triglycerides molecule and transported via the lymphatic system.

3. The stomach is capable of limited absorption of short chain fatty acids. These are present mainly as a result of the action of gastric lipase on foods that contain high amounts of short chain fatty acids.

4. The major by products of lipids digestion are long chain free fatty acids and monoglycerides. These products are absorbed from micelles formed during fat digestion and resynthesized into new triglycerides in the villi.

5. The triglycerides are then combined with cholesterol, protein, phospholipids and other substances and covered with a protein coat. This collective structure of fat and protein is called as lipoprotein. This particular type of lipoprotein contain approximately 85 per cent triglycerides& specifically called as chylomicron. Chylomicron are the principal lipoproteins made in the intestinal wall. This chylomicron enters the lymphatic system and eventually the bloodstream through thoracic duct to carry most of all absorbed fat from the diet to target tissues for energy production or for storage as fat in muscle, adipose or liver tissue.

In the blood stream, chylomicron will be acted upon by lipoprotein lipase (LPL). LPL de-esterify (breakdown the triglycerides in chylomicron into free fatty acid and glycerol. Much of glycerol travel via blood to cells in the liver or kidney where it can be made into glucose.

6.6.3 Transport between Body Tissues

The level of lipids in the blood stream is not only determined by those absorbed by the intestine but also by those released from the liver. The liver can take fatty acid directly from the blood stream or synthesize them from other nutrients such as glucose, amino acids or alcohol. These fatty acids are combined with glycerol to form triglycerides. The triglycerides are packaged with cholesterol, phospholipids and protein before being released into the bloodstream. These protein packets are known as lipoprotein and are necessary to allow lipids to move through the water medium of the blood. There are three major lipoproteins in addition to the chylomicrons:

1. Very low density lipoprotein (VLDL): consist mainly of triglycerides that are being transported to provide fatty acid& glycerol to tissues.

2. Low density lipoproteins (LDL): contain a high proportion of cholesterol, LDL transport cholesterol from the liver to other body cells. They are called bad cholesterol because they may be taken up by muscle cells in arteries and are involved in atherosclerosis.

3. High density lipoproteins (HDL): are dense because they are made up of a large population of protein. They also contain a high proportion of cholesterol; However HDL transport cholesterol in the opposite direction from LDL. HDL removes cholesterol from the peripheral tissues and return it to the liver for possible degradation. This function prevents atherosclerosis and coronary heart diseases. So, they are referred as good cholesterol.

A general composition of VLDL, LDL and HDL is mentioned below in Table 6.5.

Table 6.5: Compostion of VLDL, LDL and HDL in TGL, PL and others

	Chylomicron	*VLDL*	*LDL*	*HDL*
TGL	85 per cent	50 per cent	10 per cent	4 per cent
Phospholipid	9 per cent	18 per cent	20 per cent	24 per cent
Cholesterolester	3 per cent	12 per cent	37 per cent	15 per cent
Protein	2 per cent	10 per cent	23 per cent	55 per cent
Cholesterol	1 per cent	7 per cent	8 per cent	2 per cent
Others	–	3 per cent	–	–

Questions

Q. 1. How can fatty acids be classified?

Q. 2. How the dietary lipid can be digested in the body?

Q. 3. Explain the roles phospholipids and sterol in the body?

References

1. Wardlaw's, Perspectives in Nutrition, 8th Edition, McGraw-Hill Companies, ISBN 978–0–07–296999–3.

2. Nutrition by David C. Nieman, Diane E. Butterworth and Catherine N. Nieman, 1990 edition, Wm. C. Brown Publisher.

3. http://en.wikipedia.org/wiki/Hypocholesterolemia

4. http://library.med.utah.edu/NetBiochem/fattyacids

7
Vitamins

7.0 Introduction

There is a saying with respect to vitamins "if a little is good, then more must be better." But our total vitamin needs to prevent deficiency conditions are relatively less. Plants can synthesize all the vitamins they require; animals vary in their ability to synthesize vitamins. Generally, humans require about 28 g (1 oz) of vitamins for every 70 kg (150 lb) of food they consume. Some people believe that consuming vitamins far in excess of their needs provides them with extra energy, protection from disease, and prolonged youth. For example, guinea pigs and humans are among the few organisms that cannot make their own supply of vitamin C. Before the discovery about vitamins, certain foods were known to cure some diseases for example during the 15th and 16th centuries, many British sailors on long sea voyages died from the disease scurvy. Later it was discovered that by eating lemons and limes prevents scurvy, so citrus fruits were included in the sailors' rations, and deaths from scurvy declined sharply. Ancient greeks, used to treat night blindness with beef liver, which was later realized as a rich source of vitamin A. In India, turmeric was very popular for its healing power and was known to impart immunity. Although vitamins were not discovered until the 20th century, vitamin A was known for more than 3500 years as a factor needed to prevent night blindness. Some vitamin deficiencies are still of public health concern in specific groups of people in developed countries and in large populations in many developing countries.

Vitamins are divided into fat-soluble vitamins and water-soluble vitamins.

Essential Vitamins For Human Being

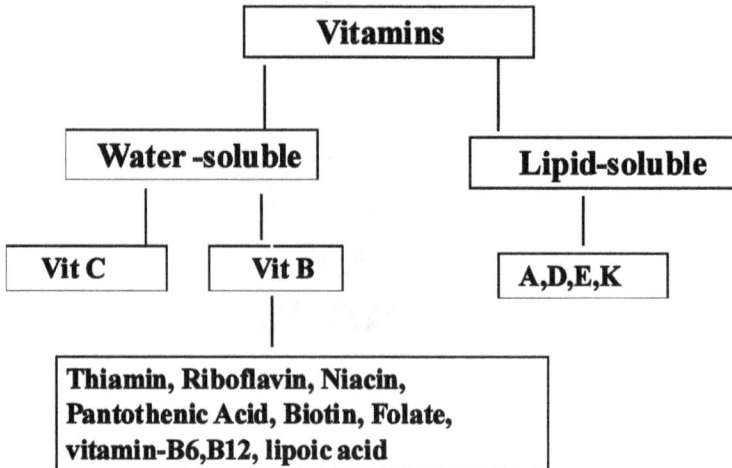

```
                          ┌─────────────────┐
                          │    Vitamins     │
                          └─────────────────┘
              ┌──────────────────────┴──────────────────────┐
    ┌──────────────────────┐              ┌──────────────────────┐
    │   Water -soluble     │              │    Lipid-soluble     │
    └──────────────────────┘              └──────────────────────┘
        ┌──────────┴──────────┐                     │
  ┌───────────┐        ┌───────────┐          ┌───────────┐
  │   Vit C   │        │   Vit B   │          │  A,D,E,K  │
  └───────────┘        └───────────┘          └───────────┘
                             │
        ┌─────────────────────────────────────────┐
        │ Thiamin, Riboflavin, Niacin,            │
        │ Pantothenic Acid, Biotin, Folate,       │
        │ vitamin-B6,B12, lipoic acid             │
        └─────────────────────────────────────────┘
```

7.1 Are Vitamins Essential in Diet?

We don't eat vitamins for obtaining energy; even then they are essential as they support growth, development, and maintenance of body tissues. Vitamins are organic substances (containing carbon bonded to hydrogen) required in small amounts in the diet. Vitamins are indispensable in human diets because they either cannot be synthesized in the body at all or are synthesized in inadequate quantities. However, a substance does not qualify as a vitamin merely because the body can't make it. Scientists have identified 13 vitamins as essential. Vitamins were named alphabetically: A, B, C, D, and E but later, some substances originally classified as B-vitamins were removed from the list as they were shown to be non-essential substances. The B-vitamins originally were thought to have a single chemical form but turned out to exist in many forms. Thus, "Vitamin B" now comprises 8 B-vitamins. Vitamins are classified as Fat soluble and water soluble vitamins. Fat Soluble vitamins include - Vitamins A, D, E, and K as they dissolve in organic solvents, such as ether and benzene. Water-soluble vitamins dissolve in water and include B-vitamins and vitamin C.

Our body can receive required amount vitamins from foods (of plant and animal origin) as diet and from dietary supplements. But it is imperative to consume enough vitamins (in the forms the body can use) to prevent the occurrence of deficiency disease. Whether vitamins are supplied from the diet or from vitamin supplements, they are usually similar chemical compounds and work equally well in the body. However, vitamins consumed in foods as part of a diet may be more beneficial than vitamins taken as dietary supplements.

Deficiency disease appears when the vitamin intake is insufficient to meet body's requirement this may lead to decline in health or poor health condition. The deficiency and related symptoms can be lessened by increased intakes of the vitamin but this will be applicable when the deficiency is not in advanced stages. Along with correcting deficiency diseases, a few vitamins have been useful as pharmacological agents as drugs in treating several non-deficiency conditions. But these treatments often require the administration of megadoses (amounts much higher than typical human needs for the vitamin). For example, megadoses of a form of niacin can be used as part of blood cholesterol–lowering treatment for certain individuals.

Absorption of Vitamins

Water soluble vitamins like B-vitamins and vitamin C can be absorbed in the small intestine independent of dietary fat. Absorption of water-soluble vitamins typically ranges from 90 to 100 per cent.

Fat-soluble vitamins like Vitamins A, D, E and K are absorbed along with dietary fat. Thus, adequate absorption of fat soluble vitamins depends on the efficient use of bile and pancreatic lipase in the small intestine to digest dietary fat and the adequate absorptive ability of the intestinal mucosa (Fig.7.1). Under optimal conditions, about 40 to 90 per cent of the fat-soluble vitamins are absorbed when they're consumed in recommended amounts.

How are Vitamins Transported in Body?

Once absorbed, vitamins follow mentioned below path:

1. **Water soluble vitamins** are delivered directly to the bloodstream and are circulated throughout the body.

2. **Fat soluble vitamins** are transported in manner similar to dietary fats. Vitamins are packed for transport via the lymphatic system and are delivered by the bloodstream to target cells by way of chylomicrons and other blood lipoproteins throughout the body. As chylomicrons circulate, much of its triglyceride content is removed by body cells and the remnant is taken up by the liver (these remains contain the fat-soluble vitamins absorbed from the diet). The liver then "re- packages" fat-soluble vitamins with new proteins for transport in the blood, or it stores them in adipose tissue or the liver for future use.

Malabsorption of Vitamins

Vitamins consumed in diet must be absorbed well from the small intestine to meet body's requirement. A person must eat larger amounts of vitamins in conditions when the absorption of a vitamin is reduced (condition known as malabsorption) to avoid appearance of deficiency symptoms. There could be many conditions which can lead to the vitamin malabsorption and the individuals with malabsorption are usually administered vitamin supplements to prevent deficiencies. Malabsorption can occur because of:

   ☆ GI tract and pancreatic disease which cause fat malabsorption also leads to poor absorption of fat-soluble vitamins.

   ☆ Alcohol abuse and certain intestinal diseases leads to malabsorption of some B-vitamins.

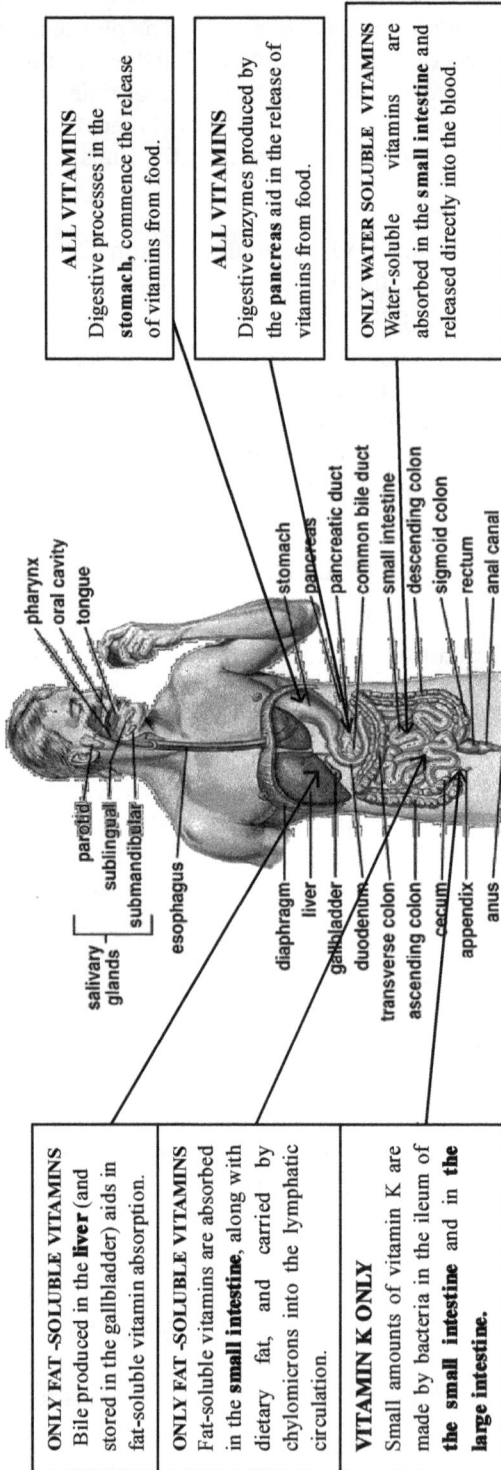

ALL VITAMINS
Digestive processes in the **stomach,** commence the release of vitamins from food.

ALL VITAMINS
Digestive enzymes produced by the **pancreas** aid in the release of vitamins from food.

ONLY WATER SOLUBLE VITAMINS
Water-soluble vitamins are absorbed in the **small intestine** and released directly into the blood.

ONLY FAT -SOLUBLE VITAMINS
Bile produced in the **liver** (and stored in the gallbladder) aids in fat-soluble vitamin absorption.

ONLY FAT -SOLUBLE VITAMINS
Fat-soluble vitamins are absorbed in the **small intestine**, along with dietary fat, and carried by chylomicrons into the lymphatic circulation.

VITAMIN K ONLY
Small amounts of vitamin K are made by bacteria in the ileum of **the small intestine** and in **the large intestine.**

Labels: pharynx, oral cavity, tongue, parotid, sublingual, submandibular, salivary glands, esophagus, stomach, pancreas, pancreatic duct, common bile duct, small intestine, descending colon, sigmoid colon, rectum, anal canal, diaphragm, liver, gallbladder, duodenum, transverse colon, ascending colon, cecum, appendix, anus

Figure 7.1: An Overview of the Vitamins Digestion and Absorption.
Source: **Perspectives in Nutrition, 8th edition, Gordon M. Wardlaw, Pal M. Insel.**

Storage of Vitamins in the Body

✰ Fat-soluble vitamins are not readily excreted from the body (as they are often stored in the liver and/or adipose tissue) with an exception of vitamin K.

✰ In contrast, most water-soluble vitamins are excreted from the body quite rapidly (exceptions are vitamin B12 and B-6, which are stored to a greater extent than the other water-soluble vitamins), resulting in limited storage in the body.

Vitamins should be consumed daily because of the limited storage of many vitamins. When the vitamin is lacking in the diet for atleast several weeks and body stores are essentially depleted then the signs and symptoms deficiency condition would appear. Thus, an occasional drop in dietary intake of most vitamins is not a serious health concern in otherwise healthy individuals.

Vitamin Toxicity

Though the toxic effects of an excessive intake of any vitamin is theoretically possible, Vitamins A and D are most likely to cause toxicity. However, these vitamins are unlikely to cause toxicities unless taken in amounts at least 5 to 10 times more than the Daily Reference Intake norm. The daily intake of balanced multivitamin and

Table 7.1: Showing the difference between Fat Soluble and Water Soluble Vitamins

Factor	Fat Soluble Vitamins	Water Soluble Vitamins
Soluble	Fat soluble vitamins are generally associated with fatty foods such as butter, cream, vegetables oils and fats of meat.	Water soluble vitamins are soluble in water.
Vitamins	Include vitamins A, D, E and K	B-vitamins and vitamin C.
Absorption	These are absorbed along with dietary fat. Under optimal conditions, about 40 to 90 per cent of the fat-soluble vitamins are absorbed when they're consumed in recommended amounts.	These can be absorbed in the small intestine independent of dietary fat. Absorption of water-soluble vitamins typically ranges from 90 to 100 per cent.
Transported in body	These are packed for transport via the lymphatic system and are delivered by the bloodstream to target cells.	These are delivered directly to the bloodstream and are circulated throughout the body.
Malabsorption	GI tract and pancreatic disease which cause fat malabsorption also lead to poor absorption of fat-soluble vitamins.	Alcohol abuse and certain intestinal diseases leads to malabsorption of some B-vitamins.
Storage	These are not readily excreted from the body with an exception of vitamin K. They are often stored in the liver and/or adipose tissue).	These vitamins are excreted from the body quite rapidly resulting in limited storage in the body. Exceptions are vitamin B12 and B-6, which are stored to a greater extent than the other water-soluble vitamins.
Toxicity	Vitamins A and D are most likely to cause toxicity	

mineral supplements usually supplies less than twice the Daily Value of the vitamin or mineral, so it is unlikely to cause toxic effects in adults.

Fat soluble and water soluble vitamins are different from each other as mentioned below in Table 7.1.

7.2 Fat Soluble Vitamins (Vitamin A, D, E and K)

7.2.1 Vitamin A

Vitamin A refers to the **preformed retinoids (active form) and pro- vitamin A carotenoids (inactive form)** that can be converted to vitamin A activity.

Retinoids is a collective term for the biologically active forms of vitamin A; Retinoids exist in three forms: retinol (an alcohol), retinal (an aldehyde), and retinoic acid (an acid). Retinoids are called preformed vitamin A because, unlike carotenoids, they do not need to be converted to become biologically active.

The tail segment of the vitamin A structure terminates in 1 of these 3 chemical groups (alcohol, aldehyde, or acid) which determines the name or classification. the synthesis of retinoic acid is a "dead end" in metabolic terms. To some extent, these forms can be interconverted (Figure 7.2).

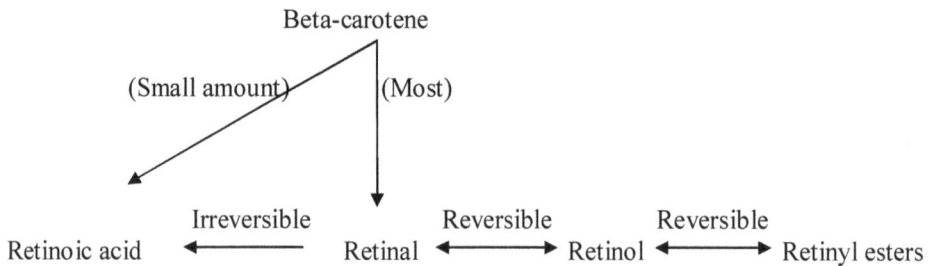

Figure 7.2: Interconversions of Beta-carotene and Various Retinoids.

The tail of the vitamin A molecule can vary from cis to trans configuration. This orientation influences the function of the specific retinoid.

Carotenoids are yellow-orange pigmented materials in fruits and vegetables, some of which are provitamins—that is, they can be converted into vitamin A. Of the 600 or more known carotenoids, only alpha-carotene, beta-carotene, and beta-cryptoxanthin can be converted to biologically active forms of vitamin A. Other carotenoids, such as lycopene, do not have vitamin A activity in humans.

Structure of Pre and Pro Vitamins are shown below in Figure 7.3.

Food Sources of Vitamin A

1. **Retinoids (preformed vitamin A)** are found in liver, fish, fish oils, fortified milk, and eggs. Margarine is fortified with vitamin A, as are fat-free, low-fat, and reduced-fat milks.

2. The **provitamin A carotenoids** are found mainly in dark green and yellow-orange vegetables and fruits, such as carrots, spinach and other greens,

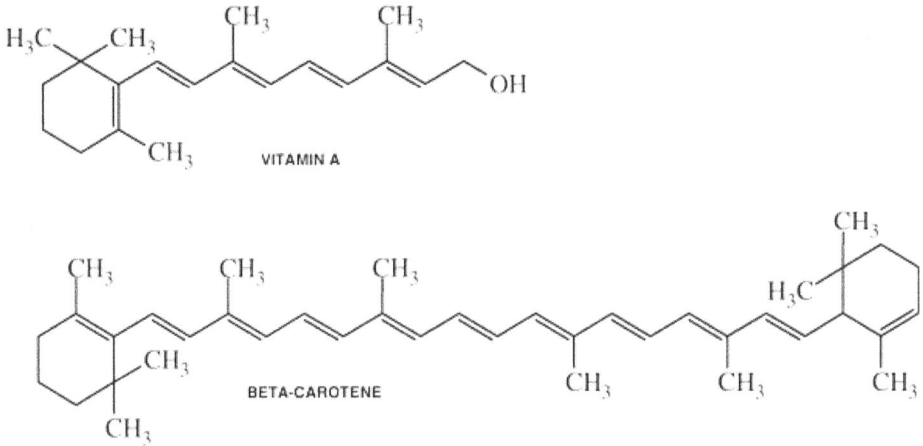

Figure 7.3: Structure of Vitamin A and β-Carotene.

winter squash, sweet potatoes, broccoli, mangoes, peaches, apricots and cantaloupe.

Beta-carotene accounts for the orange colour in carrots and other carotenoid-rich foods. In dark green vegetables, this yellow-orange colour is masked by the dark green pigment chlorophyll, although these vegetables also contain provitamin A. Therefore, consuming a varied diet rich in both dark green and yellow-orange vegetables can provide vitamin A.

Unit for Measuring Vitamin A Content

Earlier the amounts of vitamin A and most other nutrients were used to be expressed in international Units (IU's). With the advent of more sensitive means for measuring nutrient content consequently, milligram (1/1000 of a gram) and microgram (1/1,000,000 of a gram) measurements have generally replaced IU's as the units of measure. Dietary vitamin A activity is currently expressed in Retinol Activity Equivalents (RAE). Following is a tool for converting the amounts of vitamin A and carotenes expressed in 1 unit of measure into another unit of measure.

Following conversion factors can be used:

1. 1 µg of Beta-carotene= 0.125 µg Retinol Or 8 µg of Beta-carotene= 1 µg Retinol. If the diet contains both retinol and Beta-carotene, its vitamin A content can be expressed as follows:

    a. Retinol Content (µg)= µg Retinol + µg Beta-carotene X 0.125. This is applicable if retinol and Beta-carotene are expressed in µg.

Requirement of Vitamin A

Requirement of Vitamin A is showing in Table 7.2.

Table 7.2: Showing the Vitamin A Requirement

Group	Vitamin A (µg/d)	
	Retinol	Beta-carotene
Adult men and women	600 µg	4800 µg
Pregnant Woman	800 µg	6400 µg
Lactating Woman (0-12 months)	950 µg	7600 µg
Infants (6- 12 months)	350 µg	2800 µg
Childrens (1-6 years)	400 µg	3200 µg
Boys and Girls (7-18 years)	600 µg	4800 µg

Absorption, Transport, Storage and Excretion of Vitamin A

☆ Foods of animal origin contains preformed vitamin A as retinol and retinyl ester- compounds (retinol attached to a fatty acid). Retinyl esters themselves do not exhibit vitamin A activity till the retinol and fatty acid are separated in the intestinal tract.

☆ In the intestinal lumen retinol esters are hydrolysed by pancreatic lipase enzymes and brush border retinyl ester hydrolases.

☆ As the retinol is dissolved in the fat of the food so they require bile salts to emulsify them to micelles and which are then taken up by the mucosal cells of the small intestine (upto 90 per cent of retinol is absorbed into the cells of the small intestine).

☆ After absorption, when a fatty acid is attached to retinol a new retinyl ester is formed. These retinyl esters are packaged into chylomicrons before entering the lymphatic circulation.

☆ The provitamin A carotenoids can be enzymatically (dioxygenase enzyme) split within the intestinal cells or liver cells to form retinal or, to a lesser extent, retinoic acid. The carotenoid absorption is much lower than that of retinol. After being absorbed in the small intestine, carotenoids can be cleaved to yield retinal, which is then converted to retinol by a specific reductase. Retinol can then have a fatty acid attached to it to become a retinyl ester and enter the lymphatic system as part of a chylomicron. The chylomicrons transport vitamin A to tissues for storage or cellular use. Carotenoids also can enter the bloodstream directly; however, this occurs to a lesser extent.

☆ About more than 90 per cent of the body's vitamin A is stored in the liver and smaller amounts in adipose tissue, kidneys, bone marrow, testicles, and eyes.

☆ Normally, the vitamin A reserve of the liver is sufficient for several months to protect an individual against vitamin A deficiency.

☆ When vitamin A as a retinoid is released from the liver into the bloodstream, it is bound to a retinol-binding protein (RBP) in the bloodstream, retinol-binding protein is bound to another protein called transthyretin (commonly known as prealbumin). In contrast, when carotenoids are released from the liver, they are carried by the lipoprotein VLDL.

☆ Within the body's cells, retinoids are bound to specific RBPs, which direct them to the functional sites in the cell. Nearly all cells contain one or more of these binding proteins however the distribution of the cellular RBPs differs among tissues, due to their different functional requirement for vitamin A.

☆ The ester is hydrolysed and the retinol is bound to apo retinol binding protein (Apo-RBP) for transport of vitamin A to the intended tissues. The resulting RBP is secreted into the plasma where it associated with circulating prealbumin. Specific cell surface receptors are involved for the delivery of retinol from plasma to the peripheral target cells.

☆ Vitamin A is not readily excreted by the body, but some of the vitamin A is lost in the urine.

☆ Kidney disease increases the risk of vitamin A toxicity in the body as excretion of vitamin A through urinary way is affected.

Function of Vitamin A and Carotenoids

I. Functions of Vitamin A (Retinoids)

Vitamin A (retinoids) perform different functions in the body. Their key functions include growth and development, cell differentiation, vision, and immune function.

1. Growth and Development

☆ Retinoids play an important role in embryonic development.

☆ Vitamin A is also involved in the development of eyes, limbs, cardiovascular system, and nervous system. Therefore, lack of vitamin A during early stages of pregnancy may result in birth defects and fetal mortality (death).

☆ Retinoic acid is also essential for the production, structure, and normal function of epithelial cells in the lungs, the trachea, the skin, the GI tract, and many other systems. It is important for the formation and maintenance of mucous-forming cells in these organs.

2. Vision

The retina of eye consists of the rods and cones. Rods are responsible for the visual process in dim light, translating objects into black-and-white images and detecting motion. Cones are responsible for the visual process under bright light, translating objects into colourful images. Vitamin A (as retinal) is required in the retina of the eye to turn visual light into nerve signals to the brain.

    a. In the light: In the rods, ll-cis-retinal binds to a protein called opsin to form the visual pigment rhodopsin. The absorption of light will catalyzes a change in the shape of ll-cis- retinal to all-trans-retinal, causing opsin to separate from all-trans-retinal. This process of separation of protein is called as bleaching process. This leads to a series of biochemical events that bring changes in the ion permeability of the photoreceptor cells and initiate a signal to the nerve cells that communicate with the brain's visual center. Thousands of rod cells (containing millions of molecules of rhodopsin) are triggered simultaneously to produce this signal. During exposure to bright light, the rod's rhodopsin is completely activated and cannot respond to more light.

b. **In the dark:** To keep the visual process operational, all trans-retinal must eventually be converted (enzymatically) back to ll-cis-retinal. This regeneration occurs within several minutes. The ll-cis-retinal then moves back to the photoreceptor cells, where it recombines with the opsin, forming rhodopsin and is ready for another visual cycle. It is not that all retinal is used in each vision cycle, as some of the retinal is stored in the eye to maintain vitamin A pools, which is required for dark adaptation process (is the process by which the rhodopsin concentration in the eye increases in dark conditions, allowing improved vision in the dark). Dark adaptation process will be impaired in case vitamin A pools become dwindle, thus making it difficult to adjust a human to see in dim light, the condition is known as **night blindness**.

3. Cell Differentiation

Vitamin A is important for maintaining normal differentiation of the cells that make up the structural components of the eye, such as the cornea (clear lens) and the retina (rod and cone cells).

4. Immune Function

Vitamin A (mostly as retinoic acid) is important for immune system functions. As vitamin A helps in maintaining the epithelium which acts as a barrier in protecting the body against the entry of disease pathogens. So, in Vitamin A deficiency conditions there can be increased incidence of infection and individuals have greater vulnerability to illness and infection. Vitamin A supplementation has been shown to reduce the severity of some infections, such as measles and diarrhoea globally.

5. Vitamin A is Used in Dermatology

Several synthetic compounds with a chemical structure similar to that of vitamin A (called analogs) have been used in topical and oral medications to treat acne and psoriasis.

II. Functions of Carotenoid

As been discovered, that several dietary carotenoids can be converted to vitamin A within the body and carotenoids may have functions other than having provitamin A activity. The most common carotenoid is beta-carotene which has most vitamin A activity.

Diets high in carotenoid rich fruits and vegetables may decrease the risk of certain eye diseases, cancers, and cardiovascular disease. Because of its chemical structure, beta-carotene may act as an antioxidant within tissues, thereby protecting them from free radical damage. Diets containing high amounts of the carotenoids lutein and zeaxanthin may protect against age-related macular degeneration of the eye.

Deficiency Diseases of Vitamin A

An individual is at very little risk of developing vitamin A deficiency if the diet has good amount of vitamin A. However, Vitamin A deficiency is one of the major public health problems in developing countries and globally vitamin A deficiency is the leading cause of non-accidental blindness.

Vitamin A deficiency can develop in:

☆ Poor and older adults,

☆ People with alcoholism or liver disease (which limits vitamin A storage), and

☆ Individuals with severe fat malabsorption, such as gluten-sensitive enteropathy (celiac disease), chronic diarrhea, pancreatic insufficiency, Crohn's disease, cystic fibrosis, and AIDS etc.

☆ Premature infants also are at risk of deficiency because they are born with low stores of vitamin A.

Vitamin A deficiency results in many changes in the eye.

☆ When the retinol in the blood is insufficient to replace the retinal lost during the visual cycle, the rods in the retina regenerate rhodopsin more slowly. The resulting **night blindness** is a common early symptom of vitamin A deficiency.

☆ Without enough retinoic acid, mucous- forming cells deteriorate and are no longer able to synthesize mucous. The eye, especially the cornea, is adversely affected by the loss of mucous because mucous helps keep the eye surface moist and washes away dirt particles that settle on the eye. This leads to the development of **conjunctival xerosis** which refers to abnormal dryness of the conjunctiva of the eye.

☆ **Bitot's spots,** which is foamy gray spots on the eye consisting of hardened epithelial cells also appear as vitamin A deficiency worsens.

☆ Above conditions progress to **keratomalacia** refers to softening of the cornea and scarring. The sequence of changes in the eye is collectively known as **xerophthalmia** which causes irreversible blindness in millions of people globally.

☆ Vitamin A deficiency also produces skin changes, known as follicular hyper- keratosis.

☆ Keratin, is a component of the outer layers of the skin and protects the inner layers. It also reduces water loss through the skin. During severe vitamin A deficiency, keratinized cells, replace normal epithelial cells in the underlying skin layers. Hair follicles become plugged with keratin, giving a dry, rough, sandy texture to the skin.

☆ Vitamin A deficiency can impair growth in infants and young children, thus adequate vitamin A stores are required before an infant is weaned, to protect them against deficiency conditions. Vitamin A supplementation might prevent deficiency in infants and young children at risk also may protect them.

Toxicity of Vitamin A

The signs and symptoms of toxicity from excessive vitamin A in the body is called as hypervitaminosis A. This condition will occur with long-term use of vitamin supplement at 5 to 10 times the RDA for retinoids. Accordingly, the upper limit is set at 3000 µg/day of retinol to prevent harmful effects in the body. No upper limit has been set for carotenoids as vitamin A toxicity results only from excess intakes of retinoids.

Three kinds of vitamin A toxicity can appear which include: acute, chronic, and teratogenic.

1. **Acute toxicity** is caused by the ingestion of 1 very large dose of vitamin A or several large doses taken over a few days (about 100 times the RDA). The effects of this include headache, GI tract upset, blurred vision, and poor muscle coordination. Once the excess dose is stopped, the signs will disappear. An extraordinarily large dose of 500 mg in children and 10 g in adults can be fatal.

2. In **chronic toxicity** occur with repeated intakes of at least 10 times the RDA guidelines. The condition includes a wide range of signs and symptoms in infants and adults including headache, skin disorders, joint pain, loss of appetite, reduced bone minerals, liver damage, double vision, hemorrhage, and coma. The treatment involves discontinuing the supplement. The symptoms decrease in few weeks as blood concentrations fall within a normal range. Permanent damage to the liver, bones, and eyes, can occur with the chronic ingestion of excessive amounts of the vitamin.

3. **Teratogenic toxicity** is the most serious and tragic effects of hypervitaminosis A causing birth defects. Vitamin A and its related analog forms (all-trans-retinoic acid and 13-cis- retinoic acid) are used to treat various skin disorders, such as acne and psoriasis. However, these analogs are teratogenic in humans and can cause spontaneous abortion and birth defects in laboratory animals, including congenital malformations of the head (because neural crest cells, which are important in the development of the head and brain, are known to be very sensitive to excess amounts of vitamin A). Women of childbearing age should be cautioned against using these medications or should use reliable methods to prevent pregnancies that might result in fetal malformations. The FDA recommends that women of childbearing age limit their intake of preformed vitamin A to 100 per cent of the Daily Value.

Carotenoids ingested in large amounts from foods does not readily cause toxicity. The rate of conversion of carotenoids to vitamin A is relatively slow. Also the efficiency of carotenoid absorption from the small intestine decreases markedly as dietary intake increases. A condition called as hypercarotenemia, or carotenemia ("high in the bloodstream") will appear if an individual consistently consumes large amounts of carrots, carrot juice, or winter squash, that results in turning the skin colour to yellow-orange colour.

7.2.2 Vitamin D

Bone deformities that were likely caused by the vitamin D deficiency disease, rickets, have been described since ancient times. It wasn't until 1918, when scientists cured rachitic dogs (dogs affected with rickets) by feeding them cod liver oil, that diet was linked with this disease. Soon after, vitamin D was discovered and cod liver oil became a daily supplement for millions of children. Most scientists classify vitamin D as a vitamin. However, in the presence of sunlight, skin cells can synthesize a sufficient supply of vitamin D from a derivative of cholesterol. Because a dietary source is not required if synthesis is adequate to meet needs, the vita- min is more correctly classified as a "conditional" vitamin, or prohormone (a precursor of an active hormone). in the absence of UV light exposure, an adequate dietary intake of vitamin D is essential to prevent the deficiency diseases rickets and osteomalacia and to provide for cellular needs. After exposure to the sun, humans produce vitamin D_3 (cholecalciferol) from a derivative of cholesterol. The liver and kidneys each add a hydroxyl group (–OH) to this to yield the active form of vitamin D (1,25 dihydroxy D_3, or calcitriol). Figure 7.4 shows the chemical structure of Vitamin D_2 (Ergocalciferol) and D_3 (Cholecalciferol).

Sources of Vitamin D_2

a. Dietary Sources

The best food sources of vitamin D are fatty fish (*e.g.*, sardines, mackerel, and salmon), cod liver oil, fortified milk, and some fortified breakfast cereals. The Vitamin

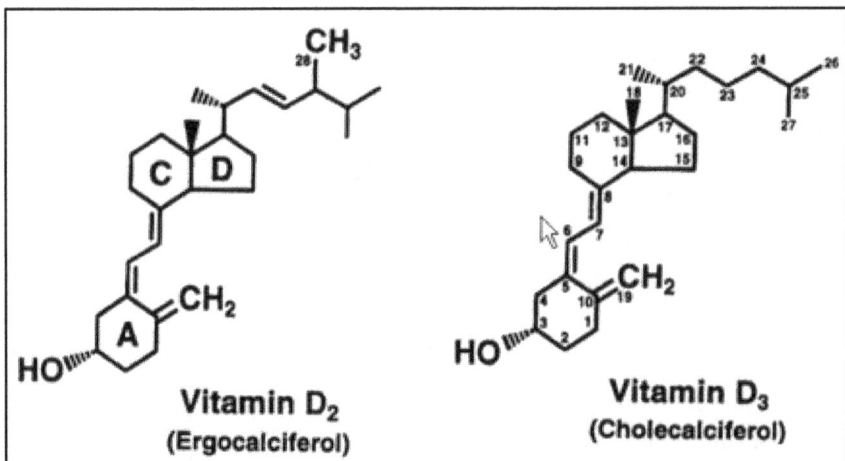

Figure 7.4: The Chemical Structure of Vitamin D$_2$ (Ergocalciferol) and D$_3$ (Cholecalciferol).

D content of food sources from animals varies with the diet, breed and exposure to sunlight of the animal.

In India, milk is generally fortified with vitamin D. Although eggs, butter, milk, and margarine are poor source of vitamin D, but fortitification can increase the vitamin D level. Most fortified foods and supplements containing vitamin D are in the form of ergocalciferol, or vitamin D_2, the same form found naturally in foods. Ergocalciferol has vitamin D activity in humans, but in lesser amounts than provided by cholecalciferol (vitamin D_3).

b. Synthesis of Vitamin D₃ in the Skin

The synthesis of vitamin D_3 begins with a compound called 7-dehydrocholesterol, a precursor of cholesterol synthesis located in the skin. During exposure to sunlight, Ultraviolet light of 290-310 nm wavelength penetrates the epidermis of skin and one ring on the molecule undergoes a chemical transformation, forming the more stable vitamin D_3 (cholecalciferol) by photolysis. This change allows vitamin D_3 to enter the bloodstream for transport to the liver and kidneys, where it undergoes hydroxylation (the addition of –OH) and subsequent conversion to its bioactive form 1,25 dihydroxy D_3 (calcitriol). For many individuals, sun exposure provides 80 to 100 per cent of the vitamin D_3 required by the body.

The amount of sun exposure needed, however, depends on the time of day, the geographic location, the season of the year, one's age, one's skin color, and the use of sunscreen. Older people are advised to get small amounts of sun exposure, especially during early morning and late afternoon (to minimize risk of skin cancer), or to take vitamin D supplements to prevent deficiency.

Using sunscreens with an SPF higher than 8, although useful in decreasing the risk of skin cancer, also may prevent adequate vitamin D_3 synthesis. The large amount of melanin (skin pigment) in dark-skinned individuals may block UV light and prevent adequate vitamin D_3 synthesis. Scientists recommend that people expose their hands, face, and arms to UV light at least 2 or 3 times a week for 10 to 15 minutes. Individuals with dark skin may need sun exposure of 30 minutes or more (or vitamin D supplementation). Prolonged sun exposure is not likely to result in vitamin D synthesis beyond needs or in toxic amounts, as excess amounts of previtamin D_3 in the skin are rapidly degraded. Persons who do not get enough UV light exposure to synthesize adequate amounts of vitamin D_3 so, they are suggested to take adequate sources of vitamin D in their diets. Figure 7.5 shows the synthesis and metabolism of vitamin D.

Vitamin D Requirement

For vitamin D there is an adequate intake level rather than a RDA value. However a more accurate RDA level could not be set because the amount of vitamin D produced by sun exposure varies considerably among individuals. The expert group of ICMR has not recommended dietary intake of Vitamin D for Indians. Only in those cases where the Vitamin D requirement is not met due to inadequate exposure to sunlight the ICMR recommends 400 µg/day of Vitamin D.

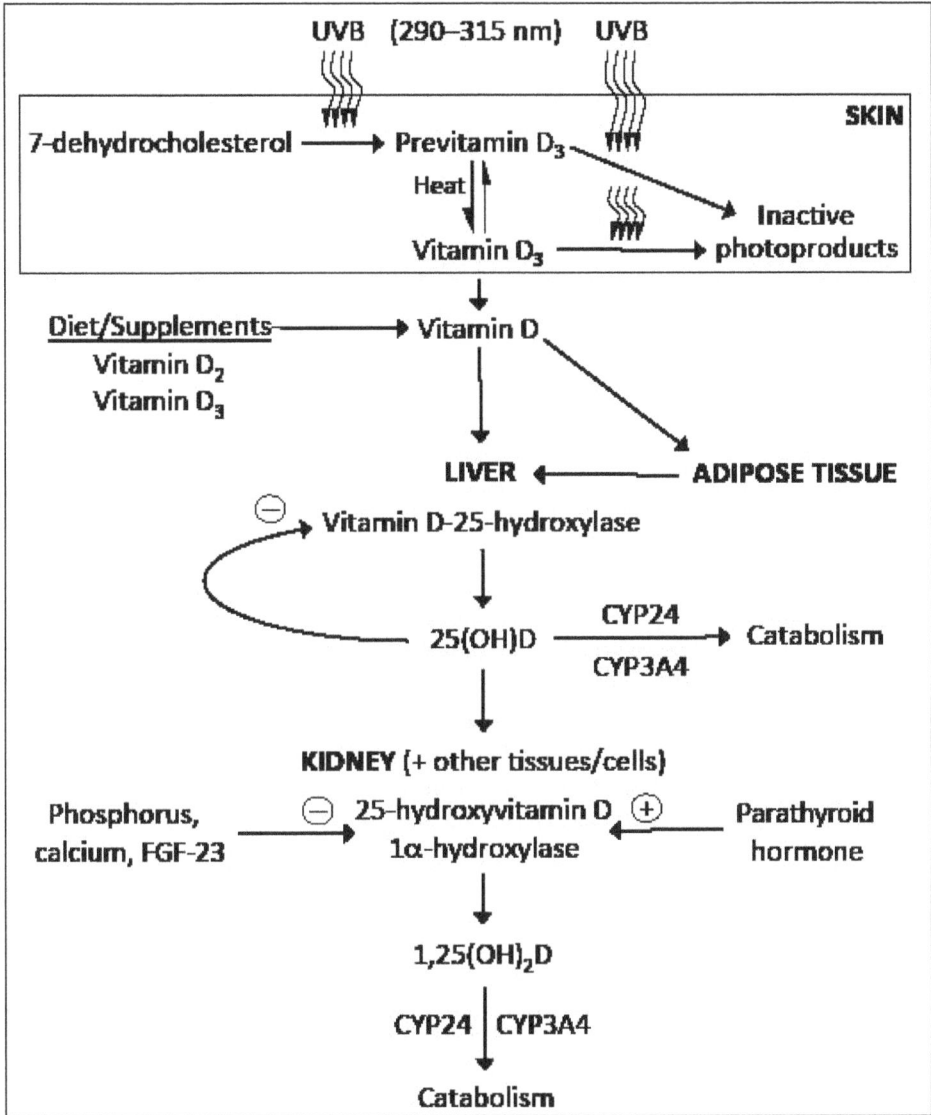

Figure 7.5: Synthesis and Metabolism of Vitamin D.

Absorption, Transport, Storage, and Excretion of Vitamin D

After consumption of vitamin D_2 containing foods, about 80 per cent of vitamin D_2 is incorporated (along with other dietary fats) into micelles in the small intestine, absorbed, and transported to the liver by chylomicrons through the lymphatic system. People with diseases that may result in fat malabsorption syndromes (like cystic fibrosis, crohn's disease and celiac disease) are at higher risk of vitamin D_2 malabsorption and deficiency.

When vitamin D (either D$_3$ synthesized in the skin or D$_2$ consumed in food or supplements) enters the body, it is bound to a protein for transport to the adipose cells for storage or to the liver and kidneys.

Upon entering in the liver, vitamin D is hydroxylated on carbon 25, converting it to 25–OH vitamin D$_3$. This inactive form circulates in the blood for many weeks. It then enters kidney, the principal site for the production of 1,25 (OH)$_2$ vitamin D$_3$, also known as calcitriol. This is the active form of the vitamin to induce vitamin D functions.

The synthesis of 1, 25(OH)$_2$ vitamin D$_3$ is regulated by the parathyroid gland and the kidneys. When there's a shortage of calcium in the blood, the parathyroid gland increases the production of parathyroid hormone (PTH). PTH then increases the production of 1, 25 (OH)$_2$ vitamin D$_3$ in the kidneys. Vitamin D is primarily excreted through the small amount of bile lost during digestion also some amounts of vitamin D is excreted in the urine.

Functions of Vitamin D

1. Vitamin D has hormone like functions, which regulate the body's concentration of calcium and phosphorus.
2. Vitamin D help in absorption of calcium and phosphorus from foods to maintain blood levels of these minerals. Thus making these minerals available to body and for their incorporation into bones. But on the other hand, when blood levels of calcium and phosphorus start to fall, vitamin D (with PTH from the parathyroid gland) can release calcium and phosphorus from bone into the blood to restore blood levels of these minerals.
3. Vitamin D maintains the calcium and phosphorous levels in the body by stimulating,
    a) Absorption in the gastro intestinal tract.
    b) Retention by the kidney
4. Vitamin D is also linked with the decreasing the risk of certain types of infections and autoimmune diseases, such as multiple sclerosis, diabetes, hypertension, and certain cancers etc.

Deficiency Diseases of Vitamin D

Vitamin D deficiency in the body leads to decreased absorption of calcium which can lead to muscular tetany, rickets in children and osteomalacia in adults.

1. Due to faulty calcification of bones, without adequate calcium and phosphorus in the blood available for deposition in the bone, the bones fail to mineralize normally. This causes the bones to weaken, and the condition primarily affects children. The disease is called rickets. The signs of rickets include enlarged head, joints, rib cage, and a deformed pelvis, bowed legs, delayed closing of frontanel, slow erruption of teeth.

    The incidence of rickets is more in children who have dark skin, children's with low milk intakes, and with negligible sun exposure due to protective clothing, sunscreens, and/or inadequate outdoor activity.

2. Vitamin D deficiency condition affecting adults is known as Osteomalacia, which means "soft bones." Osteomalacia is a condition where the quality of the bone is reduced. It occurs in women who are not exposed to sunlight and who have depleted mineral reserves resulting from successive pregnancies and prolonged lactation. Osteomalacia is associated with low phosphorous level but low blood calcium level is the most frequent cause.

Osteomalacia is also likely to occur in adults with kidney, liver, gallbladder, or intestinal disease because these diseases affect both vitamin D metabolism and calcium absorption. Other individuals at risk include those with dark skin and with limited UV exposure.

The symptoms of osteomalacia are softening of the bones, deformities of the limbs, spine, thorax and pelvis, demineralization of the bones, pain in pelvis, lower back and legs, by poor calcification of newly synthesized bone, resulting in fractures in the hip, spine, and other bones.

Toxicity of Vitamin D

Intake of excessive amounts of Vitamin D leads to toxic symptoms which include irritability, nausea, vomiting and constipation and the condition is called as **Hypervitaminosis**. Vitamin D toxicity can occur by taking excessive vitamin D supplementation. However toxicity does not result from surplus sun exposure (as vitamin D in the skin is broken down) or from natural sources in the diet.

Surplus blood calcium may leads to deposition of calcium in the kidneys, heart, and lungs; leading to anorexia, nausea, vomiting, bone demineralization, weakness, joint pain, and joint disorientation. However the symptoms are often treatable if detected in the early stages of toxicity and if excess vitamin D dose is withdrawn. Though, if excess supplementation continues, vitamin D toxicity ultimately can be fatal.

7.2.3 Vitamin E

Vitamin E is known as antisterility vitamin as it is required for normal reproduction in men and animals. The significance of vitamin E was firstly noted in 1922, when it was discovered that a substance named tocopherol (derived from the Greek words toco) present in vegetable oil was required for normal reproduction in rats. The greek word tocopherol has been taken from word "toco" meaning "childbirth" and pherein meaning "to bear."

Vitamin E has a long carbon chain tail (this tail exist in many possible isomer forms) attached to a ringed structure. Vitamin E is consists of eight naturally occurring compounds namely four tocopherols (alpha, beta, gamma, delta) and four tocotrienols (alpha, beta, gamma, delta) with widely varying degrees of biological activity. Alpha-tocopherol is the most active form. This form is present in foods and is incorporated in vitamin supplements. Gamma-tocopherol is a probable benecial form of vitamin E found in many vegetable oils. but, it does not have as much biological activity as alpha-tocopherol.

Dietary Sources of Vitamin E

Vitamin E is widely distributed in foods particularly higher concentrations are found in vegetable oils and in cereal grains. Cottonseed, canola, wheat germ, sunflower seeds, mustard oil, almonds, safflower oil, eggs, Sesame (til) oil, peanuts and asparagus are good sources. Products made from the plant oils like salad dressings, margarine, shortenings etc. are also good sources of vitamin E. Vitamin E content of a food depends on harvesting, processing, storage and cooking because it is highly vulnerable to destruction by oxygen, metals, light and deep-fat frying. Thus, foods that are highly processed and/or deep-fried are usually poor sources of vitamin E.

Meat, fruits and vegetables contain small amounts. Animal fats and dairy products contain little vitamin E. Figure 7.6 shows the structures of Vitamin E.

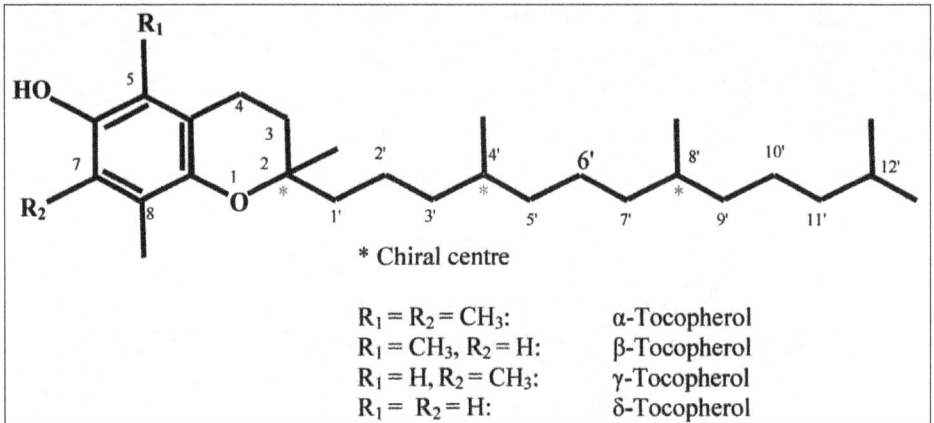

$R_1 = R_2 = CH_3$:   α-Tocopherol
$R_1 = CH_3, R_2 = H$:   β-Tocopherol
$R_1 = H, R_2 = CH_3$:   γ-Tocopherol
$R_1 = R_2 = H$:   δ-Tocopherol

Figure 7.6: Structures of Vitamin E Family.

Requirement of Vitamin E

The requirement of Vitamin E is linked to that of essential fatty acids (linoleic and linolenic acids). The requirement of Vitamin E is 0.8 mg/g of essential fatty acid.

The recommendation is based on the amount of vitamin E needed to prevent a breakdown of red blood cell membranes, a process called hemolysis. On average adults normally consume only two-thirds of the RDA for vitamin E each day from dietary sources

Absorption, Transport, Storage, and Excretion of Vitamin E

Vitamin E absorption is similar to that of other fat soluble vitamins. Digestion and absorption in the small intestine and is facilitated by presence of bile and pancreatic enzymes (hydrolyze the acetate and succinate esters of vitamin E before they can be absorbed), vitamin E must be incorporated into micelles in the small intestine.

The degree of vitamin E absorption depends on the amount of vitamin consumed and is linked to absorption of dietary fat. Vitamin E absorption from dietary intake can vary from 20 to 70 per cent. Mid gut is the site for maximal absorption.

Tocopherol, predominantly enters blood via lymph in which it is associated with chylomicrons and VLDL. Once taken up by the intestinal cells, vitamin E is incorporated into chylomicrons for transport by the lymph and eventually the blood. As chylomicrons are broken down, most of the vitamin E is carried to the liver as chylomicron remnants. A small amount is carried directly to other tissues. The liver repackages the vitamin E from the chylomicron remnants with other lipoproteins (VLDL, LDL, and HDL) for delivery to body tissues. Vitamin E is carried by these lipoproteins as it does not have a specific transport protein in the blood unlike vitamin A.

In normal human adults, serum vitamin E levels are around 1.0 mg/100ml while in new born infants the levels are low around 0.2- 0.4 mg/100ml. Vitamin E also differs from other fat-soluble vitamins in that it does not accumulate in the liver; instead adipose tissue is the site of maximal vitamin E storage.

Vitamin E can be excreted via the bile, urine, and skin. However, because vitamin E absorption is often low, most vitamin E is excreted via the small amount of bile that exits the body in the feces.

Functions of Vitamin E

1. Vitamin E is the major antioxidant *in vivo* (in the body) and the primary defence against lipid peroxidation. Thus it functions as an antioxidant that stops chain reactions caused by free radicals that can potentially damage cells. It serves to protect polyunsaturated fatty acids (PUFA) from oxidation in cells and maintain integrity of the cell membrane. It also prevents the oxidation of beta carotene and Vitamin A. Vitamin E helps to maintain cell membrane integrity and protect RBC against hemolysis.

2. Vitamin E reduces platelet aggregation

3. Vitamin E is essential for the iron metabolism and the maintenance of nervous tissues and immune function.

4. Vitamin E is known as an anti-aging vitamin, because as cells age they accumulate lipid breakdown products. Vitamin E prevents this accumulation in maintaining cell health.

5. Vitamin E inhibits the conversion of nitrites present in smoked, pickled and dried foods to nitrosamines. It thus protects the organisms against the damage likely to be caused by nitrosamines which are strong tumor promoters.

6. This vitamin also play a protective role in the prostaglandin mediated disorders such as inflammation, premenstrual syndrome and circulatory irregularities like nocturnal leg cramps and stickiness of platelets.

Deficiency of Vitamin E

Apparently vitamin E deficiency is rare to occur in humans as it is associated with the deficiency of other fat soluble vitamins. Individuals with fat malabsorption conditions (which includes cystic fibrosis and Crohn's disease), smokers and preterm infants are at greatest risk of vitamin E deficiency. Preterm infants are particularly susceptible because they are born with limited stores of vitamin E and have insufficient intestinal absorption of this vitamin. Smokers are at increased risk of deficiency because of oxidative stress caused by smoke. Following are the deficiency symptoms which may appear:

1. Prolonged intake of Vitamin E deficient diets produces uncoordinated movement, weakness and sensory disturbances.
2. It leads to reproductive failure in humans.
3. In low birth weight infants vitamin E causes premature breakdown of red blood cells leading to development of haemolytic anaemia also causes irritability and oedema. Decreased haemoglobin synthesis and fragile erythrocytes with reduced life span are the characteristic features of this syndrome and is aggravated by iron therapy.
4. Defective functioning of the retina may occur leading to permanent blindness in premature infants occurs.
5. Vitamin E deficiency is associated with decreased ability of the lymphocytes.
6. Vitamin E deficiency also can impair immune function and cause neurological changes in the spinal cord and peripheral nervous system.
7. Muscular dystrophy is also associated with vitamin E deficiency.

Vitamin E Toxicity

Although vitamin E is relatively harmless but its excessive amount can interfere with functioning of vitamin k in blood clotting. This can lead to insufficient clotting leading to a risk of hemorrhaging. These risk are of particular concern in persons taking daily aspirin or anticoagulation medications to prevent blood clots.

7.2.4 Vitamin K

Vitamin K is known as the dietary anti haemorrhagic factor because of its imperative function in blood clotting mechanism. The discovery of vitamin K focussed on its role in blood clotting. A danish researcher first discovered the association between vitamin K and blood clotting when he observed that chicks fed with a fat extracted diet developed hemorrhages. Thus, he named this new lipid-soluble factor after the danish word koagulation (which means coagulation in danish) as "Vitamin K". Vitamin K, or the quinones is the family of compounds but two are more predominant: phylloquinones (vitamin K_1) of plant origin and menaquinones (vitamin K_2) of bacterial origin. Menaquinones also are synthesized by bacteria in the human colon. A synthetic compound, called menadione (K_3) can be considered a provitamin K, because K_3 can be converted to menaquinone (K_2) in the liver by substitution of side chain. Phylloquinone, the main dietary form of the vitamin is the most biologically active form. The naturally occurring vitamin K is insoluble in water but is soluble in organic fat solvent. They are sensitive to ultra violet light and are destroyed in alkaline solutions. Figure 7.7 shows the structure of Vitamin K family.

Dietary Sources of Vitamin K

Vitamin K is relatively stable to heat processing, but it can be destroyed by exposure to light. Vitamin K is fairly distributed in nature. About 10 per cent of the vitamin k

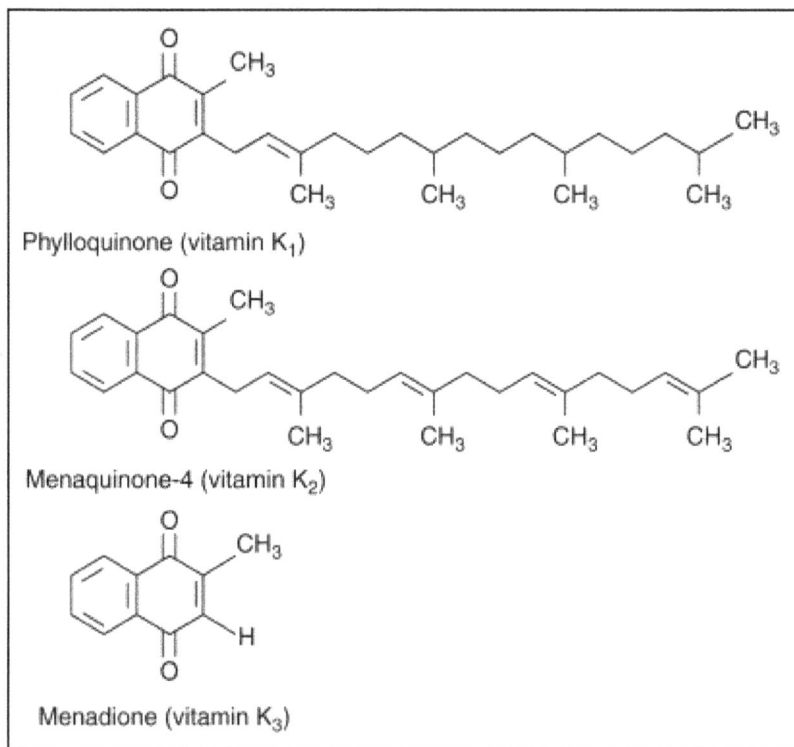

Figure 7.7: Structure of Vitamin K Family.

absorbed each day comes from bacterial synthesis in the colon. The remainder comes from dietary sources. The abundant sources of this vitamin in plant origin are those belonging to *Brassica* species, like dark green leafy vegetables (*e.g.*, spinach, parsley, kale, turnip greens, salad greens and cabbage) and tomatoes. Broccoli, peas, and green beans are also best sources of this vitamin. Vegetable oils, such as soy and canola, also are good sources. Among the animal food liver is a rich source of vitamin K. Fruits, tubers, seeds, dairy and meat products also contain Vitamin K. Vitamin K (K_2) is also synthesized by several bacteria even by those of normal mammalian intestine.

Requirement of Vitamin K

The ICMR committee considered that no recommendation is needed for this Vitamin, as the synthesis of Vitamin K occurs in the lower intestine by the colonic bacteria and present widely in foods. And a human is able to meet its daily requirement of this vitamin.

Absorption, Transport, Storage, and Excretion of Vitamin K

Dietary vitamin K is absorbed in the small intestine and the process is facilitated by the presence of bile and pancreatic juice. Approximately 80 per cent of dietary vitamin K as phylloquinone and menaquinone is taken up by the small intestine and incorporated into chylomicrons and is transported to liver via lymph. Vitamin K absorption ranges from 15-65 per cent.

The menaquinones synthesized by bacteria in the colon are absorbed, but they cater only 10 per cent of the vitamin K requirement. Menadione (K_3) is more rapidly and completely absorbed before entering portal circulation. In the liver, K_3 is first converted to K_2 by isoprenyl side chain substitution on the third carbon by mirosomes. Vitamin K is then released from liver into circulation bound to lipoproteins VLDL and finally to LDL for transport throughout the body or for storage in the liver. No specific transport protein for vitamin K in circulation has been identified so far. Liver (is the primary site), spleen, skin and muscle are the principle sites of vitamin K in humans and animals. Most of the vitamin K excretion occurs via the bile that passes out of the body in the feces, with a small amount metabolites of vitamin K_1 and K_2 are excreted via the urine.

Functions of Vitamin K

1. Synthesis of blood clotting proteins. Vitamin K is required by liver for synthesis blood clotting factor. It is essential for the activation of prothrombin. This gets converted to thrombin, which in turn activates fibrinogen to form fibrin.

   The process of blood clotting follows as below:

   Wounded tissue releases thromboplastin, which catalyses formation of prothrombin. Vitamin K catalyses the conversion of prothrombin to thrombin. This further causes conversion of fibrinogen to fibrin which forms the clot. Drugs like warfarin that strongly inhibit this reactivation process, act as powerful anticoagulants. People taking anticoagulation drugs to lessen blood clotting should maintain a consistent dietary vitamin K intake and avoid vitamin K supplementation.

2. Vitamin K also may aid to protect the body from inflammation, thus provides protection against cardiovascular disease and osteoporosis.

Deficiency of Vitamin K

Although vitamin K deficiency is very rare in normal adults or children, but can appear in patients on prolonged dose of antibiotics that disrupt vitamin K synthesis or with chronic fat malabsorption in conditions like ulceration, sprue, colitis, biliary obstruction and idiopathic steatorrhea. Vitamin K deficiency is marked by defective blood clotting and low levels of prothrombin and haemorrhage are observed during severe vitamin K deficiency.

As the new born babies have a sterile intestinal tract thus lack in the colonic bacterial colonies and due to low reserves of vitamin k at the birth so primary deficiencies arise in new borns and infants resulting in delayed and defective blood clotting and hemorrhage.

Toxicity of Vitamin K

Although vitamin K can be stored in the liver and bone in the body but it is more readily excreted than other fat-soluble vitamins. As on date, there is no upper limit set for vitamin K. Higher amounts of vitamin K natural forms (phylloquinones or menaquinones) does not cause harmful effects in the body but higher amounts of synthetic form (menadione) of vitamin k results in excess bilirubin in the blood, hemolytic aneamia, and death in newborns.

7.3. Water Soluble Vitamins (Vitamins B Complex, C)

After vitamin A (a fat soluble vitamin) the second vitamin discovered was water-soluble and was designated as "vitamin B." Although this water-soluble substance was originally thought to be a single chemical compound, but it was later researched to show that "vitamin B" actually consist of several similar compounds. Therefore numbers were added to the letter B starting from 1 to 8 to distinguish these compounds. From the discovered eight B-vitamins, only two are still referred to by letter and number and they are vitamin B-6 and vitamin B12. And the rest are now are referred to by the following names:

☆ Thiamine (formerly B-1),

☆ Riboflavin (formerly B-2),

☆ Niacin (formerly B-3),

☆ Pantothenic acid,

☆ Biotin, and

☆ Folate.

However, earlier names are sometimes used on vitamin supplement labels. Vitamin C a water-soluble vitamin was the third vitamin discovered. Figure 7.8 shows how water soluble Vitamins work to maintain overall health.

Common Classification

☆ Vitamin B family

☆ Vitamin C

Common Features of Water Soluble Vitamins

☆ Water-soluble vitamins are essential organic substances required in small amounts for the normal function, growth, and maintenance of body.

☆ These vitamins are water soluble and are generally readily excreted through urine.

☆ Unlike fat-soluble vitamins, only small amounts of water-soluble vitamins are stored in the body. So, the risk of water-soluble-vitamin toxicity tends to be low because, unlike fat-soluble vitamins, water-soluble vitamins are readily removed by the kidneys and excreted in the urine.

☆ These vitamins cannot be easy to be stored in the body, so require dietary inception.

☆ These vitamins participate in energy metabolism and function as a coenzyme. Table 7.3 lists the Vitamins B and their Coenzyme form. Some B-vitamins form more than 1 coenzyme.

Table 7.3: Lists the Vitamins B and their Coenzyme Form

Sl.No.	B-Vitamin	Coenzyme Example	Abbreviation
1.	Thiamine	Thiamine pyrophosphate	TPP
2.	Riboflavin	Flavin adenine dinucleotide Flavin mononucleotide	FAD FMN
3.	Niacin	Nicotinamide adenine dinucleotide Nicotinamide adenine dinucleotide phosphate	NAD NADP
4.	Vitamin B-6	Pyridoxal phosphate	PLP
5.	Vitamin B12	Methylcobalamin and 5-deoxyadenosylcobalamin	
6.	Biotin	N-carboxyl biotinyl lysine	
7.	Pantothenic acid	Coenzyme A	CoA
8.	Folic acid	Tetrahydro folic acid	THFA

☆ About 50-90 per cent of B vitamins are absorbed.

☆ Marginal deficiency more common.

☆ Water soluble vitamins are subjected to cooking losses. Vitamin content of any food can be decreased by contact to heat, light, air, and alkaline substances. Water-soluble vitamins can leach into cooking water, whereas fat-soluble vitamins can leach into cooking fats and oils. Retention of the vitamins B and C is greatest in foods that are prepared by steaming, microwave cooking, and stir-frying. These cooking methods limit exposure to heat and water.

☆ Fruits and vegetables are especially important sources of many vitamins.

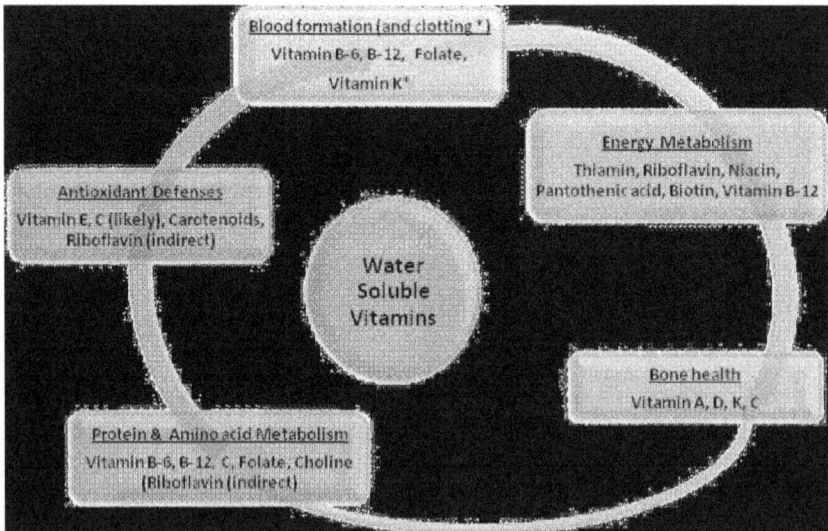

Figure 7.8: The Usefulness of Water Soluble Vitamins to Maintain Overall Health.

7.3.1 Vitamin B-1 (Thiamine)

Thiamine word is taken from *thio,* meaning sulphur, and *amine,* refer to the nitrogen groups in the molecule. Structurally thiamin consists of a central carbon atom attached to a 6-membered nitrogen containing ring and a 5-member sulphur containing ring.

The active form thiamin pyrophosphate (TPP) contains two phosphate groups to form vitamin's coenzyme. The chemical bond between each ring and the central carbon in thiamine is easily broken by prolonged contact to heat, as during cooking. This makes the vitamin no longer available for function in the body. At times baking soda (a base) is added to the cooking water of fresh green beans to retain their bright green colour and during cooking rajmah/chole to reduce cooking time but this practice is not recommended as cooking in alkaline (basic) solutions (pH e" 8.0) will also cause destruction of the bonds and make the vitamin unavailable to the body. Thiamine loss also occur during milling of grain (wheat, rice) to remove the bran and the germ, this is the reason why people on polished cereals diet suffer from thiamine deficiency as the maximum concentration of this vitamin is found in the outermost bran layer.

Chemical Nature and Properties

☆ Active form: Thiamine pyrophosphate (TPP)

☆ Contains sulfur and nitrogen group

☆ Destroyed by alkaline and heat

☆ Coenzyme for releases energy from carbohydrate and in glucose metabolism

Absorption, Transport, Storage, and Excretion of Thiamine

☆ Thiamine is absorbed mainly in the small intestine by a sodium-dependent active absorption process. It is transported mainly by RBC's in its thiamine pyrophosphate (coenzyme form).

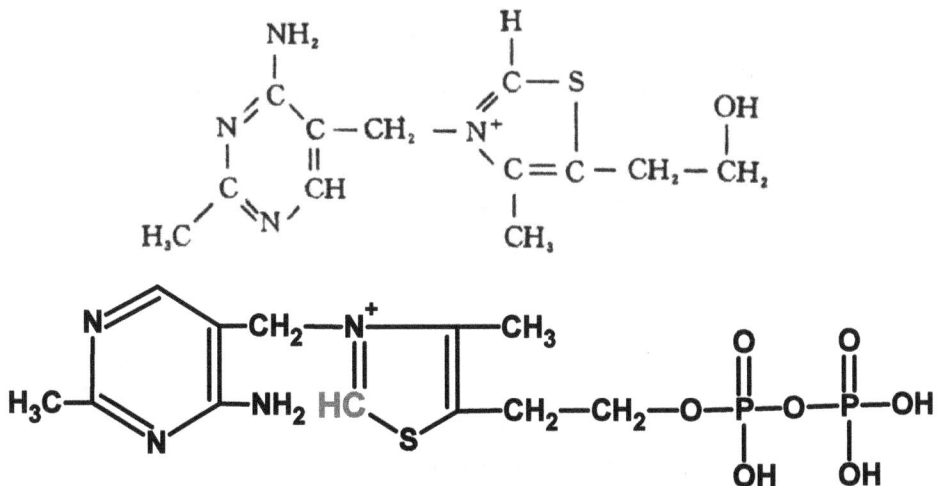

Figure 7.9: Chemical Structure of Thiamine Pyrophosphate (TPP)-Active Form.

☆ Thiamine is released by the action of phosphatase and pyro phosphatase in the upper small intestine.

☆ At low concentrations, the process is carrier-mediated, and, at higher concentrations, absorption occurs via passive diffusion. Active transport is greatest in the jejunum and ileum (it is inhibited by alcohol consumption and by folic deficiency).

☆ In thiamine absorption occurs at intakes above 5 mg/day. The cells of the intestinal mucosa have thiamine pyro phosphokinase activity, but it is unclear as to whether the enzyme is linked to active absorption.

☆ The majority of thiamine present in the intestine is in the pyro phosphorylated form TDP, but when thiamine arrives on the serosal side of the intestine it is often in the free form.

☆ The uptake of thiamine by the mucosal cell is likely coupled in some way to its phosphorylation/dephosphorylation.

☆ On the serosal side of the intestine, evidence has shown that discharge of the vitamin by those cells is dependent on Na^+-dependent ATP'ase.

☆ Very less thiamine is stored in the body as only small reserves are found in liver and muscles. Excess amount of thiamine in the body is rapidly removed out from the body by the kidneys and is excreted in the urine.

Functions of Thiamine

1. Thiamine is converted to thiamine pyrophosphate (TPP), which is an important co enzyme in the carbohydrate metabolism. TPP functions as a coenzyme for transketolase, an enzyme in the pentose phosphate pathway.

2. It is involved in transmission of nerve impulses across the cells

3. Thiamine as TPP is an essential cofactor for the conversion of amino acid tryptophan to niacin.

4. TPP is a co-enzyme of oxidative decarboxylation of α–keto acids and transketolase

5. TPP inhibit the cholinesterase activity with its effect in the nerve conduction,

Deficiency

1. The thiamine deficiency disease is known as beriberi. The word beriberi in Sinhalese (the language of Sri Lanka), means "I can't, I can't.". Beri beri is associated with the consumption of diets consisting mainly of polished cereal like white rice. Beriberi is classified as dry, wet, infantile and W.K. Syndrome.

    ☆ Individuals with beriberi are very weak because a deficiency of thiamin impairs the nervous, muscle, gastrointestinal, and cardiovascular systems.

    ☆ The symptoms of beriberi include peripheral neuropathy and weakness, muscle pain and tenderness, enlargement of the heart, oedema, difficulty breathing, anorexia, weight loss, poor memory, and confusion.

2. Peripheral Neuritis is also observed in thiamine deficiency.

   ☆ The nervous system is affected because of its reliance on glucose for energy.

3. In thiamin deficiency, glucose metabolism is severely disrupted because pyruvate cannot be converted to acetyl-CoA, the entry compound into the citric acid cycle.

Types of Beri Beri

Dry Beri Beri	Wet Beri Beri	Infantile Beri Beri	Cerebral Beriberi or Wernicke-Korsakoff Syndrome
☆ Difficulty in walking ☆ Numbness in hands and feet ☆ Loss of muscle function ☆ Mental confusion ☆ Pain and vomiting ☆ Involuntary eye movement	☆ Increased heart rate ☆ Dyspnea – shortness of breath ☆ Peripheral oedema-swelling of legs ☆ In fatal conditions it may lead to heart failure and weakening of capillary walls	It occur in first few months of life if the diet of the mother is deficient in thiamine. ☆ Symptoms are restlessness, sleeplessness, constipation, enlargement of the heart and breathlessness	☆ Found mainly among heavy users of alcohol. ☆ As alcohol decreases thiamin absorption, alcohol increases thiamin excretion in the urine, and alcoholics may consume a poor-quality diet without enough thiamin. Because thiamin is not readily stored in the body, the syndrome can occur rapidly. ☆ The symptoms include changes in vision (double vision, crossed eyes, rapid eye movements), ataxia, and impaired mental functions.

Food Sources of Thiamine

Although thiamine is found in a wide variety of foods but generally in small amounts. Foods rich in thiamin are pork products, sunflower seeds, and legumes. Whole and enriched grains and cereals, green peas, asparagus, organ meats (*e.g.*, liver), peanuts, and mushrooms and eggs also are good sources.

Wheat bran, Rice bran and wheat germ are very rich sources of thiamine while white flour is devoid of the vitamin.

Food contains some anti thiamine factors which make the thiamine unavailable.

Anti Thiamine Factors

☆ Heat Senstive **Thiaminase** present in raw fish.

☆ Heat resistant polyphenolic compounds like tannins and caffeic acid.

☆ Therefore, ingestion of tea or coffee and chewing betel nut adversely affects thiamine absorption.

RDA for Thiamine

Thiamine is involved in the carbohydrate metabolism. Its requirement is related to energy derived from carbohydrate. The ICMR expert group recommends an allowance of 0.5 mg per

1000 Kcal for adults and for infants 0.3 mg/1000 Kcal is suggested. The recommended dietary allowance per day is given in Table 7.4.

Table 7.4: The RDA of Thiamine

Group		Thiamine mg/day
Man		
	Sedentary	1.2
	Moderate	1.4
	Heavy work	1.7
Woman		
	Sedentary	1.0
	Moderate	1.1
	Heavy work	1.4
Pregnant woman		+0.2
Lactation	(0-6 months)	+0.3
Infants	0-6 months	0.2
	6- 12 months	0.3
Children	(1 – 3 years)	0.5
	(4 – 6 years)	0.7
	(7 – 9 years)	0.8
Boys	(10 – 12 years)	1.1
Girls	(10 – 12 years)	1.0

7.3.2 Vitamin B_2 (Riboflavin)

The word flavin is derived from a latin word flavous which means yellow. Active form of Vitamin B_2 : mononucleotide (FMN) and flavin adenine dinucleotide (FAD). This vitamin is sensitive to UV radiation (sunlight), therefore milk and other products good in riboflavin is stored in paper, opaque plastic containers. Chemical structure of Vitamin B-2, FMN and FAD are given in Figure 7.10.

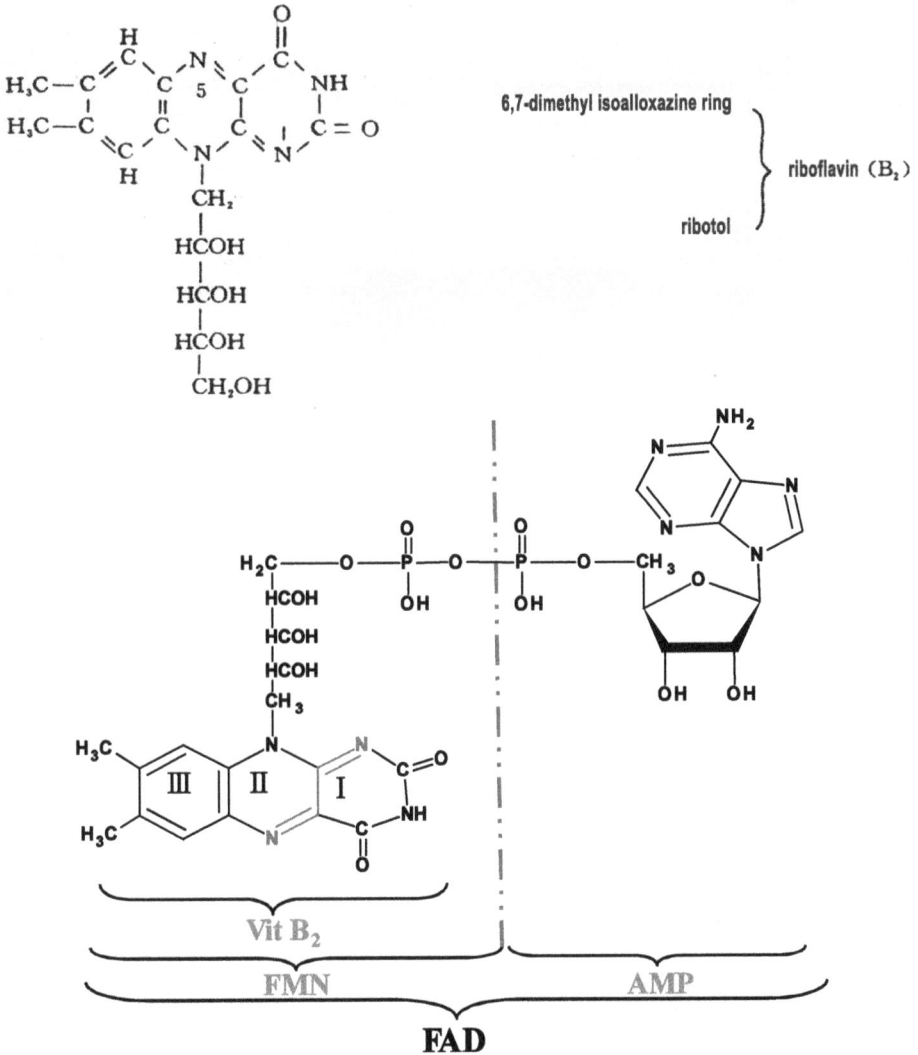

Figure 7.10: Chemical Structure of Vitamin B-2, FMN and FAD.

Absorption, Transport, Tissue Uptake and Excretion

1. Riboflavin attached non-covalently to proteins may be freed by the action of HCl secreted with in the stomach and by Gastric and Intestinal enzymatic hydrolysis of the protein.

2. Riboflavin in foods as FAD and FMN and riboflavin phosphate must also be freed prior to absorption.

3. Within the intestinal lumen FAD converted into FMN by FAD pyrophosphatase.

☆ FAD   FAD pyrophasphatase   FMN

$$\longrightarrow$$

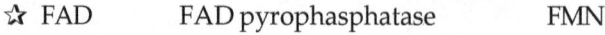

And then FMN is converted into free riboflavin

☆ FAD   FMN pyrophasphatase   FMN

$$\longrightarrow$$

☆ In case of riboflavin phosphate

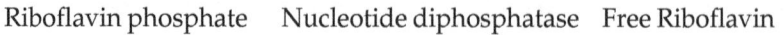

Riboflavin phosphate   Nucleotide diphosphatase   Free Riboflavin

$$\longrightarrow$$

Alkaline phosphatase

4. Free riboflavin present in foods is directly absorbed in humans by enterocytes in the upper small bowel.

5. Initial rapid uptake is by a sodium dependent and ATPase involved co-transport system.

6. The coenzymatic forms are hydrolysed to riboflavin by non-specific phosphatases and pyrophosphatases.

7. Presence of bile salts or reduction in intestinal transit increases absorption.

8. Metabolic trapping by conversion to FMN and FAD occurs before release of the vitamin into circulation by non-specific pyrophosphatases and phosphatases.

9. Riboflavin is taken up from blood by tissues in free form and not as coenzymes.

10. Liver, kidney and heart contain higher concentration of riboflavin compared to other tissues.

11. Storage of riboflavin in the blood appears to be minimal and intakes above the current requirements are rapidly excreted in the urine mainly in the form of free riboflavin. About 60-70 per cent of urinary flavin is riboflavin, the remainder comprises of other metabolites whose origin (intestinal microflora) is not much known.

Food Sources of Riboflavin

This vitamin is distributed widely in plant and animal foods and is synthesized mainly in green leaves and growing tips of plants. It is present predominantly in the free form in bovine milk but in other foods it is combined with proteins.

☆ Milk, Liver, Kidney, heart, egg white and free leafy vegetables are very good source.

☆ Major amount of vitamins in grains are present in bran and germ.

☆ Lean meat, beef, poultry, cheese, apricots and tomatoes are fair source.

☆ Cereals and legumes are not good source.

Requirement for Riboflavin

The riboflavin requirement is related to energy intake 0.6mg/1000kcal. Normally the intake is above RDA and no Toxicity has been documented. Table 7.5 lists the requirement of riboflavin for various age groups.

Table 7.5: Requirement of Riboflavin for Various Age Groups

Group		Riboflavin mg/day
Man		
	Sedentary	1.4
	Moderate	1.6
	Heavy work	2.1
Woman		
	Sedentary	1.1
	Moderate	1.3
	Heavy work	1.7
Pregnant woman		+0.3
Lactation	(0-6months)	+0.4
Infants	(0-6 months)	0.3
Children	(1 – 3 years)	0.6
Boys	(10 – 17 years)	1.3 – 1.8
Girls	(10 – 17 years)	1.0- 1.2

Functions of Vitamin B-2

This vitamin is of two types FAD – Flavin-di-nucleotide. FMN- Flavin mono-nucleotide.

FAD (Flavin-di-nucleotide) and FMN (Flavin mono-nucleotide) functions as following:

a. These substances act as coenzymes in many biological reactions primarily in oxidation – reduction, and dehydrogenation reaction. FMN and FAD are the prosthetic group of oxidoreductases with function of transmitting hydrogen.

   i) Riboflavin is important biochemically as it is vital for proper utilization of carbohydrates, fats, and proteins as energy sources.

   Riboflavin participates in metabolism following conversion to coenzyme forms FMN and FAD by the enzymes flavokinase and FAD synthetase.

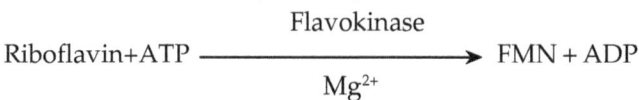

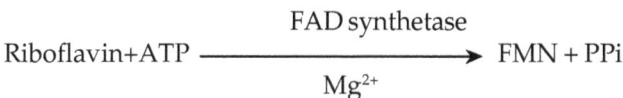

$$\text{Riboflavin+ATP} \xrightarrow[\text{Mg}^{2+}]{\text{Flavokinase}} \text{FMN + ADP}$$

$$\text{Riboflavin+ATP} \xrightarrow[\text{Mg}^{2+}]{\text{FAD synthetase}} \text{FMN + PPi}$$

      ii) Conversion of vitamin B6 and folate to active coenzymes.

c) It is essential for the formation of red blood cells.

d) It is required for the synthesis of glycogen

Biochemical Function of FMN and FAD

1. In the oxidative decarboxylation of pyruvate and alpha-ketoglutarate.

   FAD serves as an intermediate electron carrier with NADH being the final reduced product.

2. Succinate dehydrogenase is an FAD flavo protein that remove electrons from succinate to form fumarate.

3. In fatty acid oxidation "Fatty acyl CoA dehydrogenase" requires FAD as coenzyme.

4. Aldehyde oxidase using FAD converts aldehyde such as pyridoxal to pyridoxic acid and retinal to retinoic acid.

5. Synthesis of active form of folate that is 5 methyl tetrahydrofolate requires $FADH_2$.

6. Some neurotransmitters such as DOPAMINE and other other amines require FAD dependent monoamine oxidase for metabolism.

7. The synthesis of the antioxidant compound glutathione depends on the FAD containing enzyme *glutathione reductase.*

8. It is essential for normal growth and tissue maintenance.

9. It is also essential for health of eye.

Deficiency

Riboflavin deficiency is prevalent mainly among the low income groups particularly the vulnerable group and the elderly adults. Deficiency disease of vitamin B-2 is **Ariboflavinosis** (which consists of Glossitis, cheilosis, seborrheic dermatitis, stomatitis, eye disorder, throat disorder, nervous system disorder). The disease develops after 2 months on a riboflavin deficient diet usually in combination with other deficiencies and is rare in otherwise healthy people.

Riboflavin deficiency is characterized by:

1. Soreness and burning of the mouth and tongue.

2. Lesions at the angles of the mouth called **Angular Stomatitis.**

3. The inflammation of the tongue called **glossitis.**

4. Dry chapped appearance of the lip with ulcers termed **cheilosis.**

5. The skin becomes dry and results in seborehoeic dermatitis.

6. Photophobia, lacrimation, burning sensation of the eyes and visual fatigue.

7. Decreased motor co-ordination.

8. Normocytic anaemia.

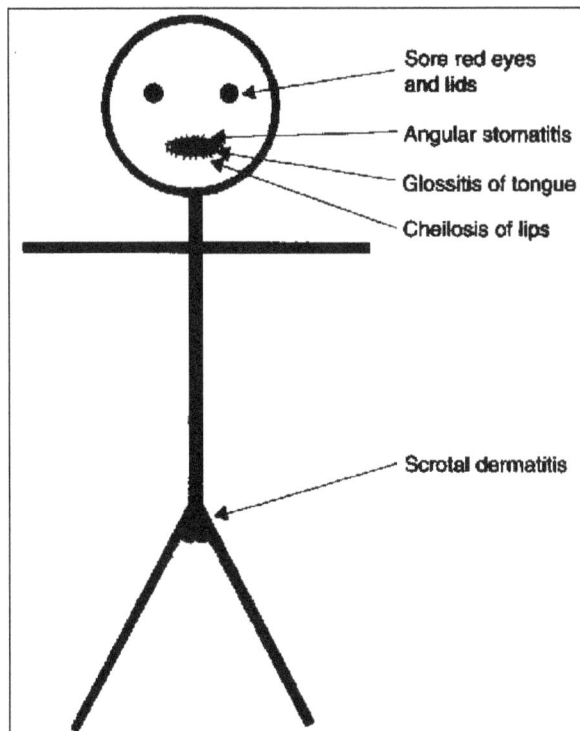

Figure 7.11: The Areas of the Human Body Affected by Ariboflavinosis.

7.3.3 Vitamin B_6 (Pyridoxine or pyridine derivatives)

Chemical Nature and Properties

Vitamin B_6 or Pyridoxine refers to related substances such as Pyridoxine, Pyridoxal and Pyridoxamine are three forms in which it is present in our body as mentioned below:

☆ Pyridoxine (PN) - Alcohol form

☆ Pyridoxal (PL) - Aldehyde form

☆ Pyridoxamine (PM) - Amine form

And their 5' Phosphates, Pyridoxine Phosphate (PNP) and Pyridoxamine-Phosphate (PMP), and Pyridoxal phosphate (PLP).

Active form of Vitamin B_6 are Pyridoxal-Phosphate and Pyridoxamine-Phosphate.

Almost all amino acids require vitamin B-6 coenzyme in their metabolism. Pyridoxine is a colourless compound which is soluble in water and alcohol. It is stable to heat, acid and alkali and but is easily destroyed on exposure to light in neutral or alkaline medium. Figure 7.12 shows the chemical structure of Vitamin B6.

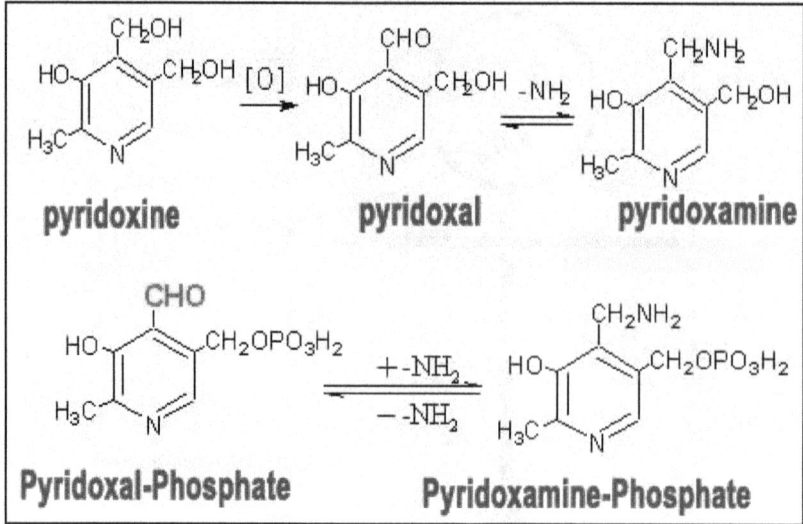

Figure 7.12: Chemical Structure of Vitamin B6 Family.

Absorption of Vitamin B6

1. For Vitamin B-6 to be absorbed, it must be dephosphorylated.

2. Intestinal alkaline phosphatases that are dependent on zinc, are involved in dephosphorylation of vitamin B6. These enzymes are found in intestinal brush border (villi).

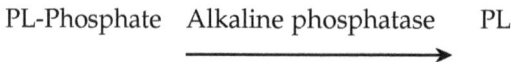

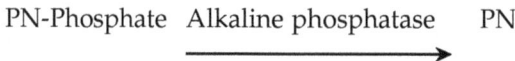

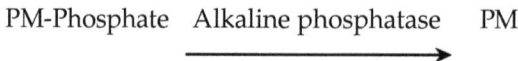

    PL-Phosphate   Alkaline phosphatase   PL

    $\xrightarrow{\hspace{3cm}}$

    PN-Phosphate   Alkaline phosphatase   PN

    $\xrightarrow{\hspace{3cm}}$

    PM-Phosphate   Alkaline phosphatase   PM

    $\xrightarrow{\hspace{3cm}}$

3. These free forms (PL, PN, PM) are absorbed by jejunum and illeum in dephosphorylated form by passive diffusion. And then it appears in liver to be converted to pyridoxal phosphate(PLP). In the plasma about 60 per cent of vitamin B-6 is PLP. Pyridoxal and PLP are primarily associated with plasma albumin.

4. At physiological intakes, the vitamin is absorbed rapidly in its free form. However, when the phosphorylated vitamers are ingested in high concentration, some of these compounds may be absorbed by itself.

5. Though this is widely distributed in the tissues, storage of vitamin B-6 in the body is very minimal. Muscle phosphorylase is the major body store (70-80 per cent) of the vitamin. Liver stores around 5-10 per cent of this vitamin, In liver the pyridoxal is converted into pyridoxal phosphate (PLP) in presence of "pyridoxal kinase".

6. The major urinary metabolites of vitamin B6 is 4-pyridoxic acid. Small quantities of other forms are also found in the urine.

Function of Pyridoxine (Vitamin B6)

1. Biochemical function Pyridoxal-Phosphate

   Vitamin B6 in the form of pyridoxal phosphate functions as a co-enzyme in many biological reactions. It acts as coenzyme of amino acid aminotransferase, decarboxylase, and d- amino-g- levulinate synthase (ALA synthase).

2. Pyridoxine is essential for the process of :

   a. **Transamination** : Transfer of amino group from one amino acid to another.

   Also involved in synthesis of nonessential amino acid via transamination.

   b. **Deamination** : Removal of the amino group. Threonine dehydratase is an enzyme that removes amino group from threonine and require PLP as coenzyme.

   c. **Decarboxylation**: Removal of the carboxyl group. The decarboxylases facilitate the synthesis of important neuroamines such as serotonin, Y-aminobutyric acid and dopamine.

2. Vitamin B6 activates enzymes needed for metabolism of CHO, fat, protein.

   a. Vitamin B6 is involved in several biochemical steps for the conversion of the amino acid tryptophan to niacin.

   b. In the carbohydrate metabolism it aids in the release of glycogen from liver and muscle. Pyridoxal phosphate is the structural component of phosphorylase which is involved in the conversion of glycogen to glucose.

3. It aids in the formation of elastin, synthesis of messenger RNA and haem part of haemoglobin. Pyridoxal phosphate dependent enzyme α-amino levulinic acid synthetase is necessary for the formation of α-amino levulinic acid which is a precusor for the biosynthesis of haem.

4. It aids in the conversion of linoleic acid to arachidonic acid.

5. Synthesize neurotransmitters.

6. Synthesize hemoglobin and WBC.

7. Pyridoxal phosphate is also involved in the biosynthesis of sphingosine from palmityl CoA and serine.

8. PLP enzyme kynureninase is involved in the tryptophan- niacin pathway.

9. Racemization which involve interconversion of D and L form of amino acid also require PLP as an co-enzyme. Although this reaction is very rare in human beings but is prevalent in case of bacteria.

10. Carnitine is a non-protein nitrogen carrying compound required for fatty acid oxidation. PLP is required for the synthesis of carnitine.

Food Sources of Vitamin B-6

Animal food contains pyridoxal and pyridoxamine phosphates as major forms while pyridoxine is found in plants and cereals. In supplements it is present as pyridoxine hydrochloride. Although vitamin B-6 in foods of animal origin is often more readily absorbed than that in foods of plant origin, whole grains are also good sources of vitamin B-6. However, vitamin B-6 is lost during the refining of grains, and it is not one of the vitamins added during enrichment. Most fruits and vegetables are not good vitamin B-6 sources, but there are some exceptions: carrots, potatoes, spinach, bananas, and avocados

1. **Well absorbed and rich sources** - Liver, Muscle meat, fish, poultry, whole grain, pulse, potatoes, milk and nuts.

2. **Less well absorbed and poor sources** - Fruits and vegetables: banana, spinach, avocado.

3. **Other Cereals are fair sources of this vitamin.**

Bioavailability of vitamin B-6 from different food sources is influenced by the food matrix and by the extent and type of processing to which the food is subjected too.

Much of the vitamin originally present in food can be lost through processing including, heating, canning, sterilization, milling.

Requirement for Vitamin B-6

The ICMR recommended dietary allowance for pyridoxine is given in table 7.6 below. Under normal circumstances average intake is more than the RDA. Athletes may require more vitamin B6. Alcohol destroys vitamin B6.

Table 7.6: The Requirement of Pyridoxin for Various Age Groups

Group		Pyridoxin mg/day
Adults		2.0
Woman		2.0
Pregnant woman and Lactating mother		2.5
Infants	(0- 6 months)	0.1
	6- 12 months)	0.4
Children	(1 – 9 years)	0.9 – 1.0
Boys and girls	(10 – 17 years)	1.6 – 2.0

Deficiency of Vitamin B6

☆ Deficiency of vitamin B6 leads to abnormalities in **protein metabolism** which is evident with poor growth, convulsions, anaemia, decreased antibody formation and skin lesions.

☆ Severe deficiency leads to **microcytic hypo chromic anaemia.**

Symptoms such as weakness, nervousness, irritability, insomnia and depression, vomiting, skin disorders, nerve irritation, impaired immune system, difficulty in walking is predominant.

Toxicity of Vitamin B6

☆ Excess of this vitamin affects the sensory nerve fibers and thus sensitivity will be destructed.

☆ Excessive dose will cause sensory and peripheral "Neuropathy" and there is loss of "Myelination".

☆ Degeneration of sensory fibers in peripheral nerves will be observed in few patients. Leading to difficulty in walking and numbness in hands/feet will be observed.

7.3.4 Folic Acid

The name of the B-vitamin folate is derived from Latin word *folium,* meaning "leaf." This is because leafy green vegetables are excellent source of this vitamin. The term folic acid refers to the synthetic form of the vitamin found in supplements and fortified foods. *Folate* is the generic name, referring to the various forms of the vitamin found naturally in foods. Folate consists of 3 parts: pteridine, para-aminobenzoic acid (PABA), and 1 or more molecules of the amino acid glutamic acid (glutamate). Chemical structure of Folic acid is given in Figure 7.13.

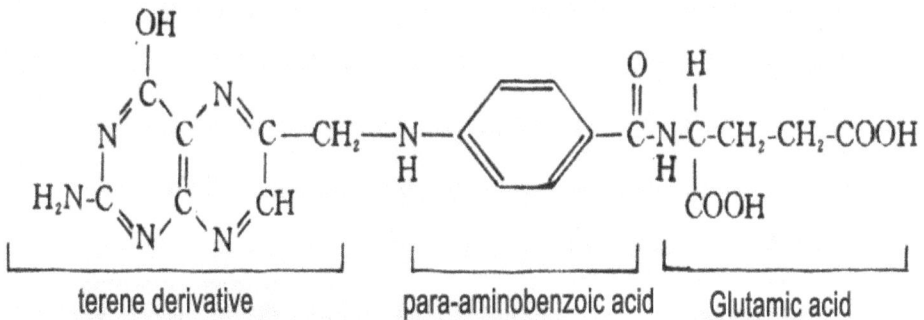

Figure 7.13: Chemical Structure of Folic Acid.

Chemical Nature and Properties

Folic acid also known as Pteroylglutamic acid and active form of folic acid is Tetrahydrofolate (FH$_4$).

☆ Folic acid is present exclusively as tetrahydrofolate derivatives in food materials and generally contains three or more glutamic acid residues in Y-glutamyl linkages to form polyglutamyl folate.

☆ It is yellowish in colour, sparingly soluble in water and readily degraded by light and UV radiation.

Absorption of Folate

☆ Folypolyglutamates are hydrolyzed to monoglutamate forms by the enzyme folate conjugase (γ-carboxypeptidase) to facilitate uptake by mucosal cells.

☆ Absorption is predominantly in duodenum and jejunum by an active process involving sodium. The absorption of folic acid from foods varies in the range of 37 to 72 per cent.

☆ Specific folate binding proteins exists in many tissues and they take part in folate transport.

☆ Folate is excreted both in urine and faeces. Feacal excretion is more than intake which shows bacterial synthesis in the intestine. Urinary excretion forms a small fraction of total excretion of this vitamin.

☆ Considerable degradation of folic acid takes place in the body. The degradative products, mainly pteridines and acetamino benzolyglutamate take longer periods for excretion in urine as compared to undegraded folate.

Food Sources of Folate

☆ Folic acid is present both in plant and animal foods. It is present in abundant amount and most bioavailable in are liver, legumes, and green leafy vegetables.

☆ Pulses, liver, egg yolk and organ meats are good sources of this vitamin.

☆ Cereal grains and milk are generally poor sources of folic acid.

☆ Folate is extremely prone to destruction by oxidation, heat, and ultraviolet light.

☆ Vitamin C in foods helps protect to folate from oxidative destruction. So, regular consumption of fresh or lightly cooked fruits and vegetables can help you maintain folate level and proper function thereof.

Biochemical Function of Folic Acid

1. Folic acid participates in one carbon transfer reactions after undergoing reduction to tetrahydrofolic acid. Dihydrofolate reductase converts folic acid first to dehydrofolate and then to tetrahydrofolate using NADPH as cofactors. Inhibitors of this enzyme such as amethopterin and aminopterin are efficient in treatment of leukemia.

2. The one carbon unit associated with tetrahydrofolate participates in biosynthesis of purines and pyrimidines and inter-conversion of amino acids.

3. Folic acid with vitamin B12 is required for the metabolism of homocysteine. Increased level of this amino acid is an independent risk factor for coronary artery disease.

4. Folate is also required by rapidly growing cells for purine and pyrimidine nucleotide synthesis, which are building blocks for RNA and DNA synthesis.

5. Neurotransmitter formation

6. Clinical application is that this vitamin can work as antitumor drug.

Deficiency of Folate

1. Deficiency disease of folate is known as **Megaloblastic anaemia.** In this there is enlargement in the size of RBC *i.e.* RBC's are lesser in number but are large in size. In this disease, bone marrow cells and other rapidly dividing cells become megaloblastic and hysegmentation of WBC occurs. Along with lesser blood cell count. RBC's are fewer than normal and are large and immature.

   ☆ Target group for this anaemia are pregnant women and alcoholics.

   ☆ Symptoms include weakness, irritability, headache, difficulty in concentration, shortness of breath.

2. Without folate, DNA damages and destroys many of the RBC's that cannot carry oxygen and they are not able to travel through the capillaries as efficiently as normal RBC's. Also there is increase in the RNA content. RNA quantity becomes greater than normal leading to excess production of other cytoplasmic constituents including haemoglobin.

3. Signs and symptoms of folate deficiency are similar of vitamin B12 deficiency.

4. Folic acid is very important during pregnancy in first two months. It is required for closure of neural tube (the embryonic tissue that forms the brain and spinal cord) and its deficiency can cause **neural tube defects or spina bifida**. **Neural Tube** Neural tube of featus closes in first 28 days of pregnancy. By the time pregnancy is confirmed, damage is done. Neural tube defect affects formation of brain and spinal cord.

5. Folate and heart diseaes: Folates play key role in breakdown of homocysteine. Without foloate homocysteine accumulate and hence formation of blood clot and atherosclerotic lesions will be observed.

Requirement of Dietary Folate

Average intake of dietary folate is normally below RDA also its excess can mask vitamin B12 deficiency. Therefore an upper limit is set at 1 mg/day. RDA for dietary Folate is given below:

   ☆ 200 µg/day for adults men, women and boys and girls (16-17 years).

   ☆ 500 µg/day for pregnant women.

   ☆ 300 µg/day for lactating mother.

   ☆ 25 µg/day for infants.

7.3.5 Vitamin B12 (Cyanocobalamin)

Pernicious anaemia was considered to be a fatal disease of unidentified origin with an unknown cure until 1926. But in the year 1926; Minot and Murphy found that pernicious anaemia could be treated by feeding patient with atleast 0.3 kg/day of raw liver. Also in 1926 Castle noted that patients with pernicious anaemia had a low level of gastric secretion. Later he suggested that the anti-pernicious anaemia factor has two components: an 'extrinsic factor' found in food and an 'intrinsic factor' found within normal gastric secretions. The extrinsic factor is now known as vitamin B12 – cobalamine and this is unique among the vitamins. Vitamin B12 is a complex molecule containing a tetrapyrrole structure with a central cobalt atom. Cyano, chloro, bromo, sulfo or nitro groups may be attached to the cobalt atom depending on the method of preparation. Vitamin B12 has a complex, multi-ring structure containing a mineral (cobalt) as part of its structure, which makes it unique among vitamins. The cyanocobalamin form of vitamin B12 forms 2 active coenzymes (methylcobalamin and 5-deoxyadenosylcobalamin) by replacing the cyano group with another group, (a methyl group or a hydroxyl group). Figure 7.14 shows the chemical structure of Vitamin B12.

Chemical Nature and Properties

☆ Cyanocobalamin is a dark red hygroscopic crystalline substance, soluble in water and alcohol.

☆ Redox compounds, aldehydes, ferrous salt, ascorbic acid, sunlight etc. destroy activity of Cyanocobalamin. But cooking losses are not excessive since it is stable at high temperature.

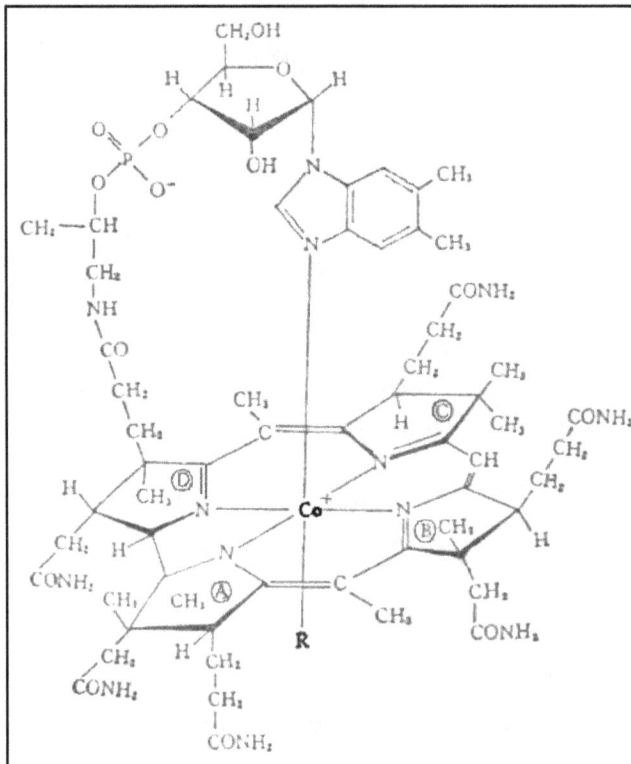

Figure 7.14: Chemical Structure of Vitamin B12.

☆ Mostly older people are at risk of vitamin B12 deficiency due to impaired vitamin B12 absorption from foods due to atrophic gastritis. Usually this deficiency is not enough to produce anaemia, but it can cause neurological problems and elevated blood homocysteine.

Digestion, Absorption, Transport and Storage of Vitamin B12

☆ Generally cobalamin is bound to proteins or polypeptides so ingested cobalamine must be released from the proteins or polypeptides to which they are linked in foods.

☆ This release usually occurs through the action of the gastric proteolytic enzyme in –pepsin and HCl in stomach.

☆ After this vitamin B12 binds to an R-Protein. This protein is present in gastric juices. The R-protein typically bind to the vitamin as it is emptied from the stomach into the duodenum region of the small intestine.

☆ R-protein are also thought to protect vitamin B12.

Absorption

☆ Within the duodenum, the R-protein is hydrolysed by the pancreatic proteases and free cobalamine is released.

☆ After released from R-protein it binds to intrinsic factor which is a Glycoprotein that carries this vitamin across intestinal mucosa and into blood stream. And this vitamin is absorbed throughout the ileum.

☆ The Vitamin B12 bound to protein is called transcobalamins, which circulates in blood. Transcobalamins exists in three forms: Transcobalamine I,II, III form. Transcobalamin II B12 complex is the major form for tissue uptake. Transcobalamine III B12 complex is taken up by hepatocytes only.

☆ Vitamin B12 is excreted via urine, bile and faeces. Biliary excretion via faeces is the major excretory route followed by this vitamin.

☆ Normally, healthy adults absorb around 50 per cent of the vitamin B12 from foods. However, the absorption of vitamin B12 can be disrupted by following defects:

❖ The absence of or defective synthesis of R-protein, pancreatic proteases, or intrinsic factor

❖ Defective binding of the intrinsic factor/vitamin B12 complex to receptor cells in the ileum

❖ The absence (or surgical removal) of much or all of the ileum and stomach

❖ Diseases in the ileum, such as Crohn's disease and Chronic malabsorption syndromes.

❖ Bacterial overgrowth in intestine and Tapeworm infestation.

Food Sources of Vitamin B12

☆ Vitamin B12 is present only in foods of animal origin which includes Liver, poultry, eggs, meat, milk and fish. Animals can make vitamin B12 from ingested soil while eating and grazing on land. Ruminant animals, such as cows and sheep, also synthesize vitamin B12 from bacteria. So, humans obtain the vitamin B12 from animal origin foods

☆ Plants do not synthesize vitamin B12 so, plant products are devoid of vitamin B12, however spirulina (blue green algae) and some seaweeds contain appreciable amounts of vitamin B12.

☆ Microbial synthesis (mainly bacteria) is the singular source of Vitamin B12 compounds in nature. Vitamin B12 is synthesized by the colonic bacteria in humans also.

☆ Yeast and most the fungi do not appear to synthesize this vitamin.

RDA for Vitamin B12

Average intake exceeds RDA. This vitamin is comparatively non-toxic (and there is no Upper Level). The recommended dietary allowance prescribed by ICMR for B12 are given in Table 7.7.

Table 7.7: Requirement of Vitamin B12

Group	Vitamin B12 mg/day
Man	1.0
Woman	1.0
Pregnancy	1.2
Lactation (0-12months)	1.5
Infants	0.2
Children (1-9 years) and Boys and girls (10-17 years)	0.2 – 1.0

Biochemical Function

Biochemical Function Methyl Form

1. Vitamin B12 has two co enzyme forms namely: deoxyadenosylcobalamin and methylcobalamine. Deoxyadenosylcobalamine acts as the coenzyme for the conversion of methymalonyl CoA to succinyl CoA in the mitochondria. Propionic acid from odd chain fatty acids and amino acids like valine and isoleucine are precursors of methylmalonyl CoA.

2. Methyl cobalamine as co-enzyme is required for the conversion of homocysteine into methionine. Methyl cobalamine transfers methyl group from 5-methyltetrahydro-folate to homocysteine to form methionine. In vitamin B12 deficiency, methly tetrahydrofolate will accumulates due to impaired transfer of methyl groups to vitamin B12.The block in the utilization of methyltetrahydrofolate in vitamin B12 deficiency results in decreased de novo synthesis of thymine leading to defective DNA synthesis, disturbances

in nuclear maturation of proliferating epithelial cells and formation of megaloblastic red blood cells.

3. Vitamin B12 is necessary for normal growth and maintenance of healthy nervous tissue and normal blood formation.

4. Vitamin B12 is involved in DNA synthesis and thus in cell replication.

5. In the bone marrow the Vitamin B12 co-enzymes are essential for the formation of red blood cells. It facilitates the formation of folate co-enzymes needed for nucleic acid synthesis.

6. Vitamin B12 is also required for the synthesis of myelin sheath that surrounds the nerve fiber.

Deficiency of Vitamin B12

1. **Pernicious anaemia**

   ☆ Deficiency is usually (95 per cent) due to decreased absorption of vitamin B12. This can be due to two reasons:

   ☆ Lack of HCl acid: as we require HCl to separate B12 from protein bound form. Without HCl the vitamin is not released from the dietary proteins, so it is not available for binding with the intrinsic factors, without the intrinsic factor the vitamin cannot be absorbed.

   ☆ Lack of intrinsic factor: Vitamin B12 deficiency is common among the elderly (50 years and above), many people especially older than 50 develops "Atropic gastritis", this condition damages the cells of the stomach. Atropic gastritis may also develop in response to iron deficiency or infection with *Helicobactor pyroli*. The bacteria implicated in ulcer formation.

The vitamin deficiency caused by lack of intrinsic factor and atropic gastritis is known as pernicious anaemia.

1. Other symptoms due to vitamin B12 deficiency are:

   ☆ Marginal vitamin B12 deficiency impairs COGNITION (loss of memory).

   ☆ Nerve degeneration, weakness

   ☆ Tingling/numbness in the extremities (parasthesia)

   ☆ Advanced neurological symptoms include a creeping paralysis that begins at the extremities (Hands and Feets) and it works inward and up the spine and can even lead to death.

   ☆ Vitamin B12 deficiency looks like folate deficiency

   ☆ There are lower number of platelets and leucocytes leading to low immunity.

   ☆ Prolonged bleeding, abdominal discomfort, neurological disorders like depression, permanent nerve damage can occur.

☆ The more chances of deficiency symptoms are more in gastrointestinal tract, bone marrow, hair follicle and nervous system. This is because they contains the cells in dividing stage.

☆ **Achlorhydria** appear especially in elderly. For treatment i**njection** of vitamin B12 is required.

☆ Takes ~20 years on a deficient diet to see nerve destruction

☆ Megaloblastic anaemia, nerve disease, High blood level of homocysteine

☆ Deficiecy will lead to progressive demyelination.

7.3.6 Vitamin C (Ascorbic Acid)

The chemical name for Vitamin C is ascorbic acid. In the year 1747, British physician Lind discovered that citrus fruit juices prevented and cured scurvy. It is well known that most animals are able to synthesize vitamin C; however humans, few birds, other primates, guinea pigs and fish are unable to synthesize vitamin C and to prevent scurvy they are required to obtain required vitamin C from dietary source.

Chemical Nature and Properties

☆ Vitamin C is also known as ascorbic acid and work as an electron donor in many processes in the human body. The term vitamin C not only refers to ascorbic acid but also to its oxidized form, dehydroascorbic acid.

☆ Ascorbic acid is highly soluble in water and is easily oxidizable in the presence of oxygen.

☆ Oxidisability is enhanced by the presence of copper, iron or alkaline pH.

☆ Ascorbic acid is white crystalline and 6 carbon compound.

☆ Its absorption decrease with high intakes and excess amount is excreted.

Absorption of Vitamin-C

☆ The vitamin C is readily available from fruits and vegetables.

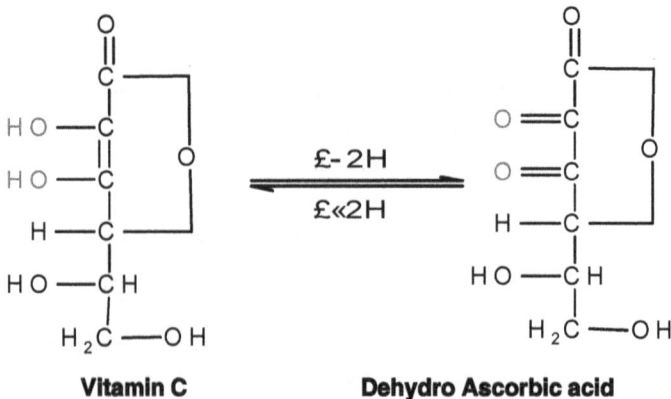

Vitamin C   **Dehydro Ascorbic acid**

Figure 7.14: Chemical Structure of Vitamin B12.

☆ It is absorbed by both active transport and simple diffusion (at a lower rate).

☆ Ascorbic acid is absorbed in the body by a sodium dependent active transport system in the small intestine. Sodium dependent active transport is required for absorption of the vitamin C, the two transporters includes-Sodium Ascorbate Co-Transporters (SVCTs) and Hexose transporters (GLUTs).

☆ Dehydroascorbic acid is the preferred form of the vitamin to be transported by diffusion to erythrocytes and leucocytes. Dehydroascorbic acid can be reduced in the body again to ascorbic acid.

☆ A significant amount of this vitamin is metabolized to oxalic acid and 2-3 diketogulonic acid.

☆ Excessive intake of the vitamin leads to excretion in urine either unchanged or as dehydroascorbic acid. Ascorbate 3-sulfate is also excreted in the urine. Other metabolites found in urine are ascorbate-2-sulfate, 2-0-methylascorbate and 2-ketoascorbitol.

Food Sources of Vitamin C

☆ Ascorbic acid is present abundantly in plant foods especially in fresh fruits and green leafy vegetables.

☆ Amla (Indian gooseberry) and guava are very rich sources of ascorbic acid.

☆ Citrus fruits, drumstick leaves, cashew fruit, agathi, cabbage, bitter gourd, oranges, tomatoes are good sources of ascorbic acid.

☆ Meat and milk contain only small quantities.

☆ Cereals and pulses are poor sources or can be practically devoid of the vitamin.

☆ Vitamin C content of pulses increases on germination.

☆ Storage conditions like temperature of storage, extent of cellular tissue damage, presence of enzyme ascorbic acid oxidase and prolonged cooking can result in considerable losses of the vitamin.

☆ Excess heat over a prolonged period, air exposure to cut surfaces, large amounts of cooking water, and exposure to baking soda reduces the level of vitamin C.

Amla is One of the Richest Source of Vitamin C

Amla can be grown abundantly throughout india and appreciable quantities can be obtained in the winter months. The fresh amla juice contains 20 times as Vitamin C than orange juice. A single amla fruits is equivalent to one or two oranges in vitamin C content.

Although heating, drying and other processing methods can leads to destruction of most or all of the vitamin C originally present in the food. However, amla as fruit is an exception not only because of its high content of vitamin C but also because of it contain substances which partially protect the vitamin from destruction on heating or drying. Amla juice which is highly acidic protects vitamin C.

Biochemical Functions of Vitamin C

1. Ascorbic Acid is essential for formation of cement substances and collagen (fibrous protein) which is found in blood vessels teeth and bones. Collagen is major structural protein which holds together the various structures of the body and gives strength to connective tissue. Collagen fibers are critical for the structure of bone and blood vessels, and they are essential in wound healing. A collagen molecule is similar to a 3-stranded rope *i.e.* it consists of 3-polypeptide chains bound together to form a triple helix. Vitamin C is required to maintain the 3 strands in the right shape to form the triple helix.

2. It is required for normal wound healing because it helps in the formation of connective tissue.

3. Ascorbic acid acts mainly as reducing agent in some hydroxylation reactions. The enzymes that require ascorbic acid for maximum activity are proline hydroxylase, lysine hydroxylase, dopamine 3-monooxygenase, peptidyl glycine-amidating monoxygenase, 4-hydroxyphenyl-pyruvate dioxygenase, trimethyllysine-2-oxoglutarate dioxygenase, and õ-butyrobetain 2-oxoglutarate 4-dioxygenase.

4. Vitamin C helps convert the structure of 2 amino acids (lysine and proline) in collagen to hydroxylysine and hydroxyproline

5. It helps in the biosynthesis of non-essential amino acids for example hydroxy, proline, tyrosin.

6. It is required for absorption of iron as it reduces ferric to ferrous form which is easily absorbed.

7. Vitamin C is required for carnitine synthesis which aids in the transport of fatty acids in the cell.

8. Vitamin C is essential for the synthesis of norepinephrine a neurotransmitter.

9. It activates hormones for example growth hormone, gastrin releasing peptide, calcitonin, gastrin, oxytocin.

10. Human body requires vitamin C for its optimal activity of drugs detoxifying metabolic systems.

11. Vitamin C is a reducing agent (antioxidant). It is an excellent anti-oxidant. It combines with free radicals oxidizing them to harmless substances that can be excreted.

12. Vitamin C is involved in Immune functions of the body. White blood cells are part of the body's immune defenses, as they contain the highest vitamin C concentration of all body constituents. This may protect against the oxidative damage associated with cellular respiration. Vitamin C may reduce this self-destruction through its antioxidant defense actions. For example, it does not prevent colds, but may reduce duration of symptoms by a day.

13. Ascorbic acid participates in the synthesis of adrenal hormones and vasoactive amines, in the metabolism of tyrosine, folate and various drugs in microsomes and in some functions of leucocytes.

14. It is also considered as nitrosation inhibitor and nitrite scavanger.

15. Ascorbic acid also stimulate the absorption of non-heam iron.

16. Eating vitamin rich food is associated with lower incidence of oesophageal and stomach cancers.

Requirement of Vitamin C

This vitamin is fairly nontoxic (at <1 gm) but the body is saturated at intake of 200 mg/day. However the upper level is 2 g/day. The recommended dietary allowances of ICMR for ascorbic acid is as given in Table 7.8.

Table 7.8: Requirement of Ascorbic Acid.

Group	Ascorbic Acid mg/day
Adult Man and Woman, Children (1-9 years), Boys and Girls (10-17 years)	40
Smokers	+35
Pregnant women	60
Lactation (0-12 months)	80
Infants (0-12 months)	25

Deficiency Disease of Vitamin C

1. Deficiency disease is called as **Scurvy**.

   Prolonged deficiency of ascorbic acid produces a disease condition called as 'scurvy' in both infants and adults. The deficiency usually appear when the diet is deficient in ascorbic acid for **20-40 days.**

   i. **Infantile scurvy**: There is loss of appetite, failure to gain weight, irritability, defective growth of bones. Haemorrhage occurs under the skin. There is defective formation of teeth and gums are swollen. The ends of the ribs become prominent resulting in beaded appearance called scorbutic rosary.

   ii. **Adult Scurvy:**

   1. General manifestation are fever, susceptibility to infection, and delayed wound healing.

   2. Anaemia: **Microcytic hypochromic anaemia** develops due to failure of absorption of iron.

   3. Gums become spongy and bleed easily. Gums become swollen and ulcerated.

   4. The blood vessels become fragile and porous due to defective formation of collagen. Joints become swollen and tender.

Table 7.9: Summary

Major Vitamin	Functions	Deciency Symptoms	Sources
Preformed (retinoids) and provitamin A (carotenoids)	Vision, cell differentiation, immunity, bone growth, reproduction	Night blindness, Poor growth, xerophthalmia, hyperkeratosis, impaired immune function	**Preformed vitamin A** (retinoids): liver, fortified milk, fish liver oils; **Provitamin A** (carotenoids): red, orange, yellow fruits & vegetables, dark green
Vitamin D	Maintenance of calcium and phosphorus concentrations, immune function	Rickets in children, osteomalacia in older adults	Milk
Vitamin E (Tocopherols Tocotrienols)	Antioxidant activity, prevention of free radicals reaction	Hemolysis of RBC's	Plant oils, seeds, nuts
Vitamin K (Phylloquinone, Menaquinone)	Synthesis of blood-clotting factors	Hemorrhage due to poor blood clotting	Intestinal synthesis by microorganisms, green vegetables
Thiamine	Coenzyme in carbohydrate metabolism and energy release	Beriberi, peripheral neuropathy, Wernicke-Korsakoff syndrome	Organ meats (e.g., liver), Pork products, sunflower seeds, legumes, whole & enriched grains and cereals, green peas, asparagus, peanuts, mushrooms and eggs
Riboflavin	Coenzyme in numerous oxidation-reduction reactions	Ariboflavinosis	Milk, Liver, Kidney, heart, egg white and free leafy vegetables
Vitamin B-6 (pyridoxine)	Coenzyme in amino acid and lipid metabolism, heme synthesis, homocysteine metabolism	Dermatitis, anaemia, convulsions, depression, confusion	Liver, muscle meat, fish, poultry, whole grain, pulse, potatoes, milk and nuts
Folate	Coenzyme in DNA synthesis	Megaloblastic (macrocytic) anaemia, birth defects	Liver, legumes, and green leafy vegetables.
Vitamin B12 (cobalamin)	Coenzyme affecting folate metabolism, homocysteine metabolism	Megaloblastic (macrocytic) anaemia, paresthesia, pernicious anaemia	Liver, poultry, eggs, meat, milk and fish, Microbial synthesis
Vitamin C (ascorbic acid)	Collagen synthesis, some antioxidant capability, hormone and neurotransmitter synthesis	Scurvy: poor wound healing, pinpoint hemorrhages, bleeding gums	Amla, Citrus fruits, drumstick leaves, cashew fruit, agathi, cabbage, bitter gourd, oranges, tomatoes

5. Clinical symptoms appear when total body pool of ascorbic acid decreases. Skin becomes rough and dry. There are small petechial hemorrhages around hair follicles.

iii. Rebound Scurvy: This occur with Sudden halt to high levels of vitamin C supplements

2. Ascorbic acid deficiency affects collagen synthesis which will affect wound healing.

3. Deficiency will lead to defective tooth formation, rupture of capillaries and impaired function of fibroblasts and osteoblasts.

4. Hyperkeratosis which include blockage of hair follicule is also associated with Vitamin C deficiency.

Excess and Toxicity of Vitamin C

Large amount of vitamin C leads to gastrointestinal distress including abdominal pain and diarrhoea. Large doses of vitamin C leads to increased risk of kidney stone. Vitamin C is metabolised in body as oxalate. Calcium oxalate is a common constituent of kidney stone.

7.3.7 Vitamin Like Substances

Vitamins are essential organic micronutrients which must be supplied in the diet. However, there are many substances which are not vitamins but function as vitamins, They are called as vitamin like substances.

The vitamin-like compounds are necessary to maintain normal metabolism in the body. They can be synthesized by the body, but their biosynthesis often occurs at the expense of other nutrients, such as essential amino acids. The need for these compounds often increases during times of rapid tissue growth, as in preterm infants. Deficiencies of these vitamin-like compounds do not exist in the average healthy adult. These substances include lipoic acid, choline, inositol, carnitine and taurine

Questions?

Q-1. What are 3 foods that are rich sources of vitamin A, D, E, K?

Q-2. How does vitamin K help in the formation of blood clots?

Q-3. What are 3 functions of vitamin D?

Q-4. What are the consequences of vitamin C deciency?

Q-5. What is the difference between rickets and osteomalacia?

Q-6. Why is the carotenoid beta-carotene classified as a provitamin?

Q-7. Why is folic acid important during pregnancy?

Q-8. Can be any deficiency of vitamin B12?

References

1. Wardlaw's, Perspectives in Nutrition, 8th Edition, McGraw-Hill Companies, ISBN 978–0–07–296999–3.

2. Srilakshmi, B. (2003), Food Science, New Age International (P) Publishers Ltd., Chennai.

3. Shakuntala Manay, M. and Shadaksharaswamy, M., (1987), Foods-Facts and Principles, New Age International (P) Publishers Ltd., Chennai.

4. Gopalan.C, B.V. Rama Sastic and Balsosubramaniam. S.C., 2012, reprinted Nutritive Value of Indian Foods. NIN, Hyderabad.

5. Bamji S. Mehtab *et al.,* Textbook of Human Nutrition, 3rd edition, 2010, Oxford and IBH Publishing Co. Pvt. Ltd., New Delhi. ISBN-978-81-204-1742-7.

6. Textbook on Nutrition and Dietetics for Higher Secondary, Text Book Corporation, College Road, Chennai. Government of Tamil Nadu, First Edition – 2004

7. 2010 revised Recommended Dietary Allowances suggested by Indian Council of Medical Research.

8. www.eatright.org

9. www.complementarynutrition.org

10. www.micronutrient.org/home.asp

11. https://online.epocrates.com/u/2911641/Vitamin+D+deficiency

12. http://www.medicaljournals.se/acta/content/files/web/3325-web-images/3325fig1_opt.jpeg

13. http://www.palmnutraceuticals.com/vite.htm

14. www.google.com

8

Minerals

8.0 Introduction

☆ Until the mid nineteenth century, the importance of minerals and vitamins was not known. The major nutrients namely carbohydrate, fat, protein were unable for promoting and sustaining growth. Hence scientists researched for the "missing elements", which are essential for growth and maintenance and they named them as minerals and vitamins.

☆ Minerals are indispensable inorganic elements necessary in small amounts in the diet for the normal function, growth, and maintenance of body tissues. Minerals are imperative for health because these cannot be synthesized in the body. These inorganic substances are critical to many body functions, including cell metabolism, nerve impulse transmission, and growth and development.

☆ Around 92 chemical elements are found on earth and atleast 50 are found in the human body. Of the nearly 45 dietary nutrients known to be necessary for human life, 17 are minerals.

☆ Deficiencies of certain minerals like calcium, iron and iodine are of major public health concern in India.

☆ If the diet lacks in the adequate level of mineral or if the minerals have been lost due to processing then, the level of the lost or lacking mineral can be increased by enrichment and or by fortification.

☆ Essential minerals which are inorganic substances are classified as macro and micronutrients based on the amount needed by humans per day.

Outline of Minerals

Macro or Major minerals are those which are vital to health and that are required in the diet by more than 100 mg or more per day. The essential macro-minerals are Calcium, Phosphorous, Magnesium, Sulphur, Potassium, Sodium and Chloride.

Micro or Trace minerals or trace minerals are those which are required in the diet in amount less than 20mg per day. Though the amount is too less but these nutrients are very important. Important microminerals of relevance in human nutrition are Iron, Zinc, Copper, Selenium, Cobalt, Fluoride, Manganese, Chromium, Iodine and Molybdenum.

Of these RDA has been established for only three namely iron, zinc and iodine. For six from the remaining seven (*i.e.* all except cobalt) are considered as "safe and adequate".

☆ Although minerals elements are required only in very small proportion to human body weight, but they play very important roles in all functions of the body. Minerals play very important role in important body processes as summarized in Figure 8.1.

☆ During growing stages (as for growing infants and children) additional amounts of some minerals is essential to ensure adequate growth of tissues.

☆ On an average, a man excretes 20-30g of mineral salts consisting of Sodium, Potassium, Magnesium, Calcium, Chloride, Sulphate and Phosphates and the lost minerals must be made available by dietary source.

☆ Humans receive their minerals requirement either from diet (from both plant and animal sources) and from dietary supplements prescribed as medicine.

☆ The quantity of minerals found in food is influenced by many agricultural factors, including genetic variability, soil, mineral composition of food, feed and medications and water.

☆ **Mineral Bioavailability** refers to the degree to which the amount of an ingested nutrient is absorbed and thus is available to the body.

☆ For some minerals, animal-based foods are the richest source and have the best bioavailability. For example, dairy products are rich sources of bio-available calcium, whereas meat is a rich source of bio-available iron and zinc. On the other hand, potassium, magnesium, and manganese are more plentiful in plant-based than animal-based food products.

Factors which influence Mineral bioavailability are:

1. The bioavailability of minerals can be greatly influenced by the amount of minerals consumed because many minerals have similar molecular weights and charges (valences). For example, magnesium, calcium, iron, and copper can each exist in the 2+ valence state. These minerals can compete with each other for absorption, thereby affecting each other's bioavailability. As an example, an excess of zinc in the diet can decrease the absorption and metabolism of the mineral copper.

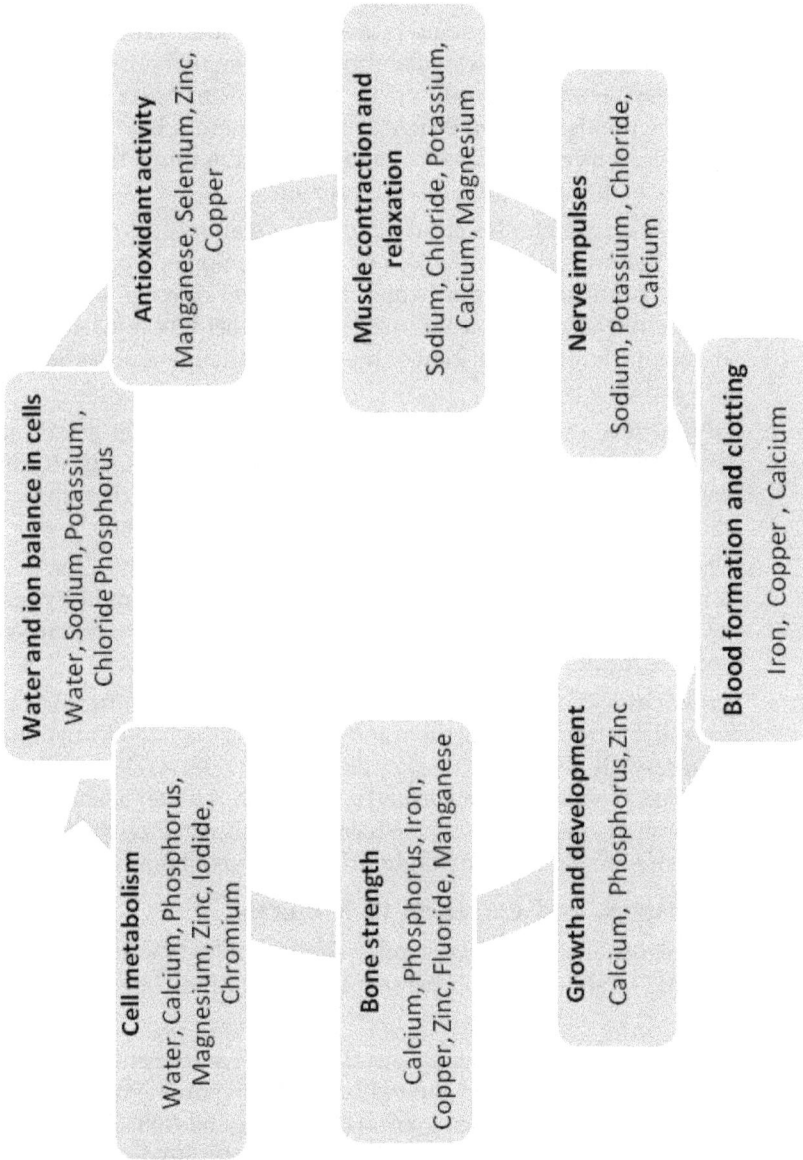

Figure 8.1: Role of Minerals in Important Body Processes.

Antioxidant activity
Manganese, Selenium, Zinc, Copper

Muscle contraction and relaxation
Sodium, Chloride, Potassium, Calcium, Magnesium

Nerve impulses
Sodium, Potassium, Chloride, Calcium

Water and ion balance in cells
Water, Sodium, Potassium, Chloride Phosphorus

Blood formation and clotting
Iron, Copper, Calcium

Cell metabolism
Water, Calcium, Phosphorus, Magnesium, Zinc, Iodide, Chromium

Bone strength
Calcium, Phosphorus, Iron, Copper, Zinc, Fluoride, Manganese

Growth and development
Calcium, Phosphorus, Zinc

2. Mineral bioavailability also is strongly affected by the presence of non-mineral substance in the diet which interfere with absorption of mineral. These substances are fibers, oxalates, polyphenolic compounds etc.

   a. The components of fiber, especially phytic acid (phytate) in wheat grain, can limit the absorption of some minerals by chemically binding to them and preventing their release during digestion. Intake of dietary fiber greatly above the adequate level of 25 to 38 g/day can adversely affect mineral status and availability. However, if grains are leavened with yeast, enzymes produced by the yeast can break some of the chemical bonds between phytic acid and minerals. Breaking of these bonds may increases the bioavailability of the minerals. Following factors can influence mineral bioavailability:

   b. Oxalic acid (oxalate) found in leafy green plants binds minerals and makes them less bioavailable. For example, Spinach contains good amount of calcium, but only about 5 per cent of it can be absorbed because of the high concentration of oxalic acid in spinach. On the other hand, about 32 per cent of the dietary calcium is absorbed from milk and milk products.

   c. Polyphenols are a group of compounds containing at least 2 ring structures that each have at least 1 hydroxyl group (OH) attached. Many polyphenols occur naturally in plants, such as tea, dark chocolate (cacao beans), and wine (grapes). Polyphenols also can lower the bioavailability of minerals, especially iron and calcium.

   d. Mineral bioavailability can be enhanced by some vitamins like vitamin C, which can improve absorption of iron when both are consumed in the same meal.

   e. Hydrochloric acid (HCl) in the stomach makes minerals more bioavailable by dissolving them and converting them to a form that can be more easily absorbed. For example, HCl provides an electron to ferric iron (Fe^{3+}) to yield ferrous iron (Fe^{2+}), which is better absorbed than ferric iron. Use of antacids and reduced stomach acid production (which is common in old age) can hinder mineral bioavailability.

Absorption, Transport, and Excretion of Minerals

Foods contain plenty amount of many minerals, but the body's ability varies to absorb and use them. The ability to absorb minerals from the diet depends on many factors:

☆ One important factor is the body's physiological requirement for a mineral at the time of consumption. In general, when the requirement for a mineral is high, such as the requirement for iron in growing children, the absorption of that mineral increases. In contrast, absorption tends to decline when the body has sufficient reserve of the mineral.

☆ Mineral bioavailability is the second important factor for the body's ability to absorb minerals.

☆ Once absorbed, minerals travel in the blood, either in a free form or bound to proteins. For example, calcium ions can be found free in the blood or bound to the blood protein albumin. Trace minerals in their free form are often highly reactive and are toxic if not bound. Thus, many trace minerals have specific binding proteins that transport them in the blood- stream. Many also are bound by specific cellular proteins once they are taken up by cells.

☆ Mineral excretion takes place primarily through the urine. However, some minerals, such as copper, are secreted by the liver into the bile for excretion in the feces. When kidney function fails, mineral intake must be controlled to avoid mineral toxicity, such as with phosphorus and magnesium.

Mineral Toxicity

☆ Excess mineral intake can be toxic, particularly trace minerals, such as iron and zinc.

8.1 Macro-Minerals

8.1.1 Calcium

Calcium is an essential element required for several life processes and is the most abundant mineral in the human body. The requirements of Calcium and Phosphorous are considered together as their function and requirement are closely linked. Over 99 per cent of the Calcium and Phosphorous is present in the bones and the remaining 1 per cent in the body fluids. The Calcium and Phosphorous are present in the ratio of 2:1 in our body.

In the skeletal system Ca and P is present in the form of hydroxyapatite crystals. Hydroxyapatite is a compound made up of calcium and phosphate that is deposited into the bone matrix to give it strength and rigidity. Calcium is an essential mineral for normal bone and tooth development. Calcium also has many industrial applications—one is as plaster of Paris, which was first used to set broken bones 1000 years ago.

Calcium Absorption, Transport, Storage, Regulation, and Excretion

☆ Calcium absorption occurs along the length of the intestinal tract. However, absorption is most efficient in the upper part of the small intestine because its slightly acidic pH helps keep the calcium dissolved in its ionic form (Ca^{2+}). Intestinal contents become more alkaline as they pass through the intestine; thus, calcium absorption decreases at the terminal end of the small intestine and colon, although some still occurs via passive diffusion.

☆ In addition, active vitamin D hormone-1, 25 $(OH)_2$ vitamin D promotes and regulates the active transport of calcium that occurs in the upper intestinal tract.

☆ When an individual has poor vitamin D status, calcium absorption is reduced. Adults absorb about 25 to 30 per cent of the calcium present in the foods they eat.

☆ Several Factors can Affect Calcium Absorption

1. During periods of growth when the body requires additional calcium such as during infancy and pregnancy stages level of calcium absorption may be as high as 75 per cent.

2. Calcium absorption is higher when calcium (as food or supplement) is ingested with other foods and not on an empty stomach.

3. Protein, other sugars, and lactose can increase calcium absorption.

4. Calcium absorption tends to decline with age (especially after age of 70 years).

5. Post- menopausal women generally absorb the least calcium.

6. Vitamin D deficiency and diarrhoea also slow down calcium absorption.

7. Other factors limiting calcium absorption include large intakes of phytic acid in fiber, oxalic acid, dietary phosphorus, and polyphenols (tannins) in tea.

8. Additionally, fat malabsorption due to some intestinal disorders can lower calcium absorption because calcium binds with fatty acids, forming unabsorbable soaps in the intestine.

☆ Calcium from the bloodstream is transported to cells either as free ionized calcium or bound to proteins. It's not just that the skeleton and teeth require more than 99 per cent of the body's calcium however, all cells have crucial requirement of calcium.

☆ Hormones regulate the concentration of calcium in the bloodstream. Even under the circumstance when calcium intake is poor, normal blood calcium can be maintained as the body withdraws calcium from the bones to keep blood and cellular calcium concentrations normal. However this condition reduces the blood calcium level. This condition leads to the release parathyroid hormone by parathyroid gland. This hormone raises blood calcium levels by working with $1,25\,(OH)_2$ vitamin D to increase the kidneys' reabsorption of calcium rather than excrete it in the urine. Parathyroid hormone also helps to increase calcium absorption indirectly by promoting the synthesis of $1,25\,(OH)_2$ vitamin D. In addition, parathyroid hormone works in conjunction with $1,25\,(OH)_2$ vitamin D to increase the calcium release from bones. When blood calcium levels rise too high, the release of parathyroid hormone falls. This causes calcium excretion via the urine to increase. The synthesis of $1,25\,(OH)_2$ vitamin D also decreases, causing a drop in calcium absorption. In addition, the thyroid gland secretes the hormone calcitonin, which blocks calcium loss from bones. All these metabolic changes keep blood calcium within the normal range. In addition to being excreted through urine and feces, dietary calcium is also excreted through the skin and feces.

Functions of Calcium

1. **Bone formation**: The major mineral ions of the bone are Calcium, Phosphorous and Magnesium. Calcium and phosphorus forms a lattice like crystal hydroxyapatite, $Ca_{10}(PO_4)_6OH_2$, which binds to the collagen fibers. This combination of materials allows bone to be resilient and strong. Collagen protein allows the skeleton to absorb impact (bone does not usually break when you jump) and the hydroxyapatite crystal makes bone strong (bone does not bend or collapse when you jump). For proper calcification of bones, (deposition of minerals on the bone matrix) which occurs during the growing years, adequate supply of these minerals is essential.

2. **Tooth formation**: Calcium and Phosphorus together as a compound is essential for the formation of dentin and enamel. The tooth consists of a hard, yellowish tissue called dentin, which is covered with enamel in the crown and cementum in the root. Enamel, the hardest substance in the body, is almost entirely hydroxyapatite crystals.

3. **Physiological Process**
   a. Calcium is essential for the clotting of blood as it is required for prothrombin activation. Because it leads to the formation of fibrin, the main protein component of a blood clot.
   b. Calcium regulates the permeability of the capillary walls and ion transport across the cell membranes.
   c. Muscle Contraction: It is essential for the contraction of the heart and skeletal muscle. When a skeletal muscle fiber is stimulated by a nerve impulse from the brain, calcium ions are released from intracellular stores within the muscle cells.
   d. Transmission of Nerve Impulses to Target Cells: When a nerve impulse reaches its target site such as a muscle, other nerve cells, or a gland the impulse is transmitted across the synapse. In many nerves, the arrival of the impulse at the target site causes calcium ions from the extracellular medium to flow into the nerve. This rise in calcium ions in the nerve triggers synaptic vesicles release their store of neurotransmitters. The released neurotransmitter then carries the impulse across the synapse to the target cells. Thus Ca regulates the excitability of the nerve fibres.
   e. Ca acts as an activator for enzymes such as rennin and pancreatic lipase.
   f. Cell Metabolism: Calcium ions help regulate metabolism in the cell by participating in the calmodulin system. Each calmodulin binds 4 calcium ions. When calcium enters a cell (often because of hormone action) and binds to the protein calmodulin, the resulting calcium-calmodulin complex activates many intracellular enzymes, including one that initiates the breakdown of glycogen.

Food Sources of Calcium

☆ Among cereals ragi contains large amounts of calcium. Bengalgram whole, gingely seeds, cuminseeds, poppy seeds, agathi, amaranth, drumstick leaves are good sources of calcium.

☆ Milk and milk products are good sources of calcium and phosphorous.

☆ Only 20 – 30 per cent of the calcium in the diet is absorbed, which is facilitated by Vitamin- D.

☆ Both the amount of calcium per serving and the bioavailability of the calcium must be considered when selecting foods as good source of calcium.

☆ Some plant foods, such as leafy greens, much of the calcium is bound to oxalic acid and is poorly absorbed. Also the absorbed amount of calcium varies widely. For example, in the 250 mg of calcium in a bowl of spinach as little as 5 per cent (about 13 mg) of calcium is bioavailable. In contrast, from 300 mg of calcium in a cup of milk nearly 100 mg is absorbable for absorption.

Requirements

ICMR recommended dietary allowances for Calcium are given in Table 8.1.

Table 8.1: Requirement of Calcium

Age Group	RDA for Calcium mg/day
Infant 0 – 12 months	500
Children 1 – 9 years	600
Adult Man and Woman	600
Boys and Girls (10-17 years)	800
Pregnant and Lactating women	1200

Deficiency of Calcium

Calcium related health problems occur due to inadequate intake, improper absorption or utilization of calcium.

1. **Osteoporosis:** Osteoporosis is the disease most often linked to low dietary intakes (including supplements) of calcium. When calcium intake is insufficient, the body will start withdrawing calcium from bone. This action is required to preserve the indispensable functions of calcium, such as those that keep the heart and muscles working. Osteoporosis is a condition found primarily among middle aged and elderly woman, where the bone mass of the skeleton is diminished. It is a condition of multiple origin. It can results due to the following reasons:

   (i) Prolonged dietary inadequacy,

   (ii) Poor absorption and utilization of calcium,

   (iii) Immobility,

(iv) Decreased levels of oestrogen in post menopausal women,

(v) Hyper parathyroidism,

(vi) Vitamin – D deficiency

2. **Osteopenia** : Osteoporosis develops over a period of many years. A failure to maintain adequate bone mass in the body first leads to a state of osteopenia called low bone mass. Osteopenia can be caused by the vitamin D deficiency disease osteomalacia, certain medications, cancer, anorexia nervosa, or other health conditions.

3. **Osteomalacia :** is a condition in which the quality but not the quantity of bone is reduced.

4. **Tetany:** is a condition of Continuous, forceful muscle contraction without relaxation. Tetany occurs when Calcium in the blood drops below the critical level. There is a change in the stimulation of nerve cells resulting in increased excitability of the nerve and uncontrolled contraction of the muscle tissue. Hence Calcium and Phosphorus ratio in the diet should be maintained at 1:1 for proper utilization of Calcium in the body.

8.2 Micro-minerals

Micro-minerals are also known as trace elements. The micro-minerals are Iron, Iodine, Zinc, Copper, Fluoride, Selenium, Chromium, Manganese, Cobalt and Molybdenum. However deficiency of just few of these elements is observed in humans. Iron and Iodine deficiencies are wide spread while deficiency of copper, zinc, chromium and selenium have been reported in recent years even in India.

8.2.1 Iron

☆ The importance of iron for the maintenance of health has been recognized for centuries. The total body iron in adults is 4g. Iron exists in a complex form in our body.

☆ It is present as :

a) Iron porphyrin compounds – Hemoglobin in RBC, myoglobin in muscle.

b) Enzymes - Peroxidases, succinase dehydrogenase and cytochrome oxidase.

c) Transport and storage forms - Transferrin and ferritin.

Absorption, Transport, Storage, and Excretion of Iron

☆ Iron is absorbed across the small intestine by a carrier-mediated mechanisms. Intestinal absorptive cells (enterocytes) produce different iron-binding carrier proteins which play an important role in the absorption process and overall regulation of iron in the body.

☆ Ferritin, a key iron-binding protein produced in the enterocyte, binds and stores mucosal iron, thereby preventing it from entering the bloodstream. The amount of mucosal ferritin produced is in proportion to body iron stores. Thus, when iron stores are low, very little ferritin is made, which allows greater amounts of iron to enter the mucosal iron pool for transport out of the enterocytes into the bloodstream. If iron stores are high, larger amounts of ferritin is made to bind iron as it enters the intestinal cells. Although a portion of this ferritin-bound iron remains in the intestinal iron pool, much of it is excreted when the intestinal cells slough off after several days. This process is called a "mucosal block" because it prevents iron from entering the bloodstream and, in effect, blocks the excess accumulation of iron. Large doses of iron, however, can overburden the mucosal block's protective ability and increase the risk of toxicity. When iron needs are high, most of the iron absorbed into enterocytes is released into an intestinal iron pool. This iron is then transported out of the enterocytes by a protein, called ferroportin, into the interstitial fluid for release into the bloodstream and distribution to body cells.

☆ In order to transport absorbed iron to body cells, the iron is oxidized from the ferrous (Fe^{2+}) form to the ferric (Fe^{3+}) form by a copper containing enzyme (either hephaestin in the enterocyte or ceruloplasmin in the blood) and bound to a serum protein called transferrin.

☆ Each transferrin molecule can bind two molecules of ferric iron for transport through the blood to body cells. All cells have transferrin receptors, located on their surface membrane, that allow them to take in the transferrin-iron complex.

☆ Cells can control the amount of iron they take in by altering the synthesis of transferrin receptors. When more iron is needed, the cell increases the number of transferrin surface receptors to enhance iron uptake. Conversely, when cellular iron requirement is low, the number of cell receptors decreases. After transferrin binds to its surface receptor, it is engulfed into the cell by endocytosis.

☆ Within the cell lysosomes, iron is released from transferrin and the receptor-protein complex is returned to the cell surface for reuse. The released iron is utilized for cellular functions or stored in the form of ferritin or hemosiderin.

☆ Because of the limited ability of body to excrete absorbed iron; intestinal absorption, cellular uptake and the storage of iron is tightly regulated. Approximately 90 per cent of the iron used each day is recovered and recycled.

☆ Only about 10 per cent is excreted, mainly some amount is lost in the form of bile in the feces. One of the proteins that aids in the regulation of iron balance in the body is called hepcidin. The loss of this protein will results in iron overload.

Food Sources of Iron

☆ The iron present in food can be as haem and non-haem iron depending upon the source from which it is obtained. Iron absorption from Indian diets is only three percent as it is mainly cereal based diet.

Table 8.2: Factors Affecting Iron Absorption

Sl.No.	Factors that Increase Absorption	Factors that Decrease Absorption
1	When the body's requirement for red blood cells is high (example during blood loss, high altitude, pregnancy, physical training).	When the body requirement for iron is low (example when high level of storage iron).
2	Lesser body reserves of iron.	Presence of phytic acid in whole grains and legumes.
3	Availability of heme iron in food.	Presence of Oxalic acid in leafy vegetables.
4	Presence of Meat protein factor (MPF).	Presence of Polyphenols as in tea, coffee etc.
5	Vitamin C intake is higher.	Reduced gastric acidity.
6	Gastric acidity favours absorption of iron.	Excessive intake of other minerals (zinc, manganese, calcium) in the diet.

☆ **Haem iron** – is obtained from animal tissues like liver, fish, poultry, meat, eggs dates are good sources of haem iron. Haem iron is absorbed and utilized better than the non- haem iron.

☆ **Non-heam iron** – is obtained from plant foods. Sources of non-haem iron are ragi, green leafy vegetables, dried fruits and jaggery.

☆ In addition to the iron in foods, iron cookware can contribute to iron intake. When foods are cooked in iron pans, small amounts of iron from the cookware are transferred to the food.

☆ Acidic food, such as tomato sauce, increases the amount of iron transferred from the cookware to the food and, in turn, increases the total iron content.

Functions of Iron

Iron plays an important role in various functions in the body. Many of these functions are dependent on iron's ability to participate in oxidation and reduction (redox) reactions, changing Fe^{2+} (ferrous) to Fe^{3+} (ferric) iron and vice versa. Although iron's ability to switch back and forth between Fe^{2+} and Fe^{3+} is very critical as this can be harmful because iron can form free radical compounds that can damage DNA and cell membranes. To prevent these destructive effects and preserve iron for healthful uses, very little free iron is found in the body. Instead, iron is tightly bound to transport, functional, or storage proteins. The chief functions of iron in the body are:

1. Iron is an essential part of two proteins, haemoglobin and myoglobin, that are involved in the transport and metabolism of oxygen. As a component of haemoglobin, iron carries oxygen in the blood from the lungs to all tissues of the body. It also transports carbon dioxide back to the lungs for expiration.

2. It forms a part of the myoglobin in muscles which makes oxygen available for muscle contraction.

3. Iron is necessary for the utilization of energy as part of the cells metabolic machinery.

4. Iron-containing enzymes play a vital role in functions such as energy metabolism, drug and alcohol transformation, and the excretion of organic compounds

Within the mitochondria, iron is a component of cytochromes that carry electrons from $NADH^+$, H^+ and $FADH_2$ to molecular oxygen in the electron transport chain. Iron also is required in the first step of the citric acid cycle for the conversion of citrate to subsequent compounds. Alcohol and many drugs are metabolized in the liver by the iron containing enzymes prior to excretion. As part of enzymes iron catalyze many important reactions in the body. Examples are

a) Conversion of beta carotene to active form of Vitamin A

b) Synthesis of carnitine, purines, collagen and neuro transmiters.

c) Detoxification of drugs in the liver.

5. Iron is a cofactor for enzymes involved in the synthesis of neurotransmitters (dopamine, epinephrine, norepinephrine, and serotonin). These are important for normal early cognitive development and lifelong brain function.

6. The immune system requires iron for the production of lymphocyte and natural killer (NK) cells that help prevent infections. If iron status is low, the effectiveness of these cells is impaired and the likelihood of infections increases. Although iron deficiency increases infection risk, iron overload also can increase the incidence of infections because bacteria require iron to grow and proliferate. Thus, iron level must be maintained within a defined range to prevent deficiency as well as overload/toxicity.

Requirement of Iron

Recommended dietary Allowances as given by ICMR for Iron for various age groups is given in Table 8.3.

Table 8.3: Requirement of Iron

Group		Body Weight (kg)	Iron requirement (mg/day)
Infants			
	0-6 months	5.4	46µg/kg/d
	6-12 months	8.4	05
Children			
	1 – 3 year	12.9	09
	4 – 6 year	18.0	13
	7 – 9 year	25.1	16
Boys			
	10 – 12 year	34.3	21
	13 – 15 year	47.6	32
	16 – 17 year	55.4	28

Contd...

Table 8.3–*Contd...*

Group		Body Weight (kg)	Iron requirement (mg/day)
Girls			
	10 – 12 year	35.0	27
	13 – 15 year	46.6	27
	16 – 17 year	52.1	26
Man		60	17
Woman		55	21
Pregnancy		55	35
Lactation		55	25

Deficiency of Iron

☆ Iron deficiency is the most common trace mineral deficiency worldwide including India. In the early stages of iron deficiency, symptoms may be minimal or unapparent because the body can mobilize stores of iron from ferritin. However, mild to moderate deficiency of iron can affect the immune function and work performance.

☆ As iron deficiency progresses and stores are depleted, the lack of iron for heme and haemoglobin synthesis results in the development of iron deficiency anaemia.

☆ This impairs oxygen transport in the blood, causing fatigue and a decreased ability to perform normal activities. Iron deficiency anaemia also compromises immune function, impairs energy metabolism, and delays cognitive development.

☆ Iron deficiency anaemia is of particular concern in young children because cognitive and developmental impairments may not be reversible.

☆ Dietary iron deficiency leads to nutritional anaemia.

☆ Nutritional anaemia is defined as the condition that results from the inability of the erythropoetic tissue to maintain a normal haemoglobin concentration. Anaemia occurs when the haemoglobin level falls below 12 gm/dl in adult man and woman.

☆ During pregnancy haemoglobin level below 11 gm/dl is termed anaemia.

☆ Nutritional anaemia is the common form of anaemia affecting women in reproductive years, infants and children which is mainly due to poor intake and absorption.

☆ Iron deficiency anaemia is wide spread in our country. The prevalence varying from 45 per cent in men and 70 per cent in women and children. The major cause of anaemia in India is because of Iron and folic acid deficiency.

☆ **Nutritional anaemia is evident as :**

1. Reduced Haemoglobin level (less than 12 g/dl).

2. Defects in the structure and function of the epithelial tissues.

3. Paleness of skin and the inside of the lower eyelid is pale pink.

4. Finger nails becoming thin and flat and eventually (spoon shaped nails) koilonychias develops.

5. Progressive untreated anaemia results in cardiovascular and respiratory changes leading to cardiac failure.

☆ The general symptoms include lassitude, fatigue, breathlessness on exertion, palpitations, dizziness, sleeplessness, dimmness of vision, and increased susceptibility to infection.

Iron Overload and Toxicity

☆ Although iron deficiency is a major public health concern as iron poses a risk for toxicity.

☆ An Upper Level of 45 mg/day has been set for iron.

☆ Intakes above this level, especially from supplements and highly fortified foods, can cause nausea and vomiting, stomach irritation, diarrhea, and impaired absorption of other trace minerals.

8.2.2 Iodine

☆ Iodine (I^2) is poisonous, can be used in a water solution as a topical anti-infective agent. The iodine ion or iodide (I^-) is the form of this trace mineral that is an essential nutrient. The term iodine is sometimes used in nutrition instead of iodide but that is not correct in nutrition perspective.

☆ It is the heaviest element needed for human health and is responsible for only one prime function in the body, the synthesis of thyroid hormones. Iodine (I^2) is an essential constituent of the thyroid hormone produced by the thyroid glands. It occurs as free iodide ions (I^-) or as protein bound iodine in our body. About 15 – 23 mg of iodine is present in the adult human body.

☆ The body store of iodine is predominantly present in thyroid gland and also in salivary gland, mammary glands gastric glands and in kidneys to a certain extent.

☆ About a billion people worldwide are at risk of iodine deficiency, and approximately 20 per cent of these people have goiter. Goiter is an enlargement of the thyroid gland, which can be caused by a lack of iodide in the diet.

Absorption, Transport, Storage, and Excretion of Iodine

☆ Most of the iodine present in foods is in the form of iodide and, to a lesser extent, iodates. These forms are very efficiently absorbed in the small intestine.

☆ After absorption, most of the iodine (the general term used for the mineral) is transported to the thyroid gland.

☆ The thyroid gland actively accumulates and traps iodide from the bloodstreams to ssupport thyroid hormone synthesis.

☆ Excess iodine is excreted primarily via the kidneys into the urine.

Factor which Affect Absorption

☆ Goitrogens are substances present in foods which cause goitre. These substances react with iodine present in the food making it unavailable for absorption.

☆ Foods like cabbage, cauliflower, raddish, raw turnips, brussels sprouts, broccoli, and cassava contain goitrogens.

Requirement of Iodine

The ICMR recommended dietary allowance for Iodine is 150µg/day. The minimum intake to prevent goiter is 50-150 µg/day.

Sources of Iodine

☆ Richest source of iodine are sea foods like sea fishes, seafood, iodized salt, molasses and common salt from sea water.

☆ Some plants contain various forms of iodide, especially the leaves of plants grown near the sea.

☆ Iodine content of vegetables, fruits and cereals depends upon the iodine content of the soil in which they grow. The soil of mountaineous regions contains less iodine.

☆ Dairy products are not naturally good sources of iodine, but they often provide significant amounts because iodide is added to cattle feeds and sanitizing solutions used in dairy processing.

☆ Breads and cereals also may contribute dietary iodine if they are prepared with iodized salt and/or dough conditioners.

Function of Iodine

1. Thyroid gland accumulates and traps iodide from the bloodstream to support its hormone synthesis. The thyroid hormones thyroxine and triiodothyronine are synthesized from the amino acid tyrosine and iodide.

2. Iodine is essential for the synthesis of the thyroid hormones T3 and T4.

3. As a component of T3, iodine is involved in the regulation of many important metabolic and developmental functions. This includes the regulation of basal energy expenditure, macronutrient metabolism, growth, brain development, and organ maturation.

Iodine Deficiency Disorders (IDD)

☆ Iodine deficiency disorders, the collective name for endemic goiter and endemic cretinism, occur when dietary iodine intake is insufficient.

☆ When iodine availability decreases and plasma levels of T4 hormone drop, the pituitary gland secretes thyroid-stimulating hormone (TSH).

☆ In response to increased TSH levels, the thyroid gland enlarges to increase its efficiency at trapping iodide. The characteristic enlargement of the thyroid gland that occurs is called a goiter.

☆ Goitre occurs in people staying in hilly regions where the iodine content of water and soil is comparatively less.

☆ In India goitre is common in hilly districts of Himalaya. Goitre can be treated by administration of iodine. If treatment is given in early stages then goitre can be corrected.

☆ If a woman consumes an iodide deficient diet during her early months of pregnancy then her infant may be born with short stature and develop mental retardation.

☆ Severe iodine deficiency in children leads to hypothyroidism resulting in retarded physical and mental growth. This condition is known as cretinism.

Iodine Toxicity

☆ The Upper Level is set at 1100 µg/day for adult men and women to prevent health- related risks.

☆ Like iodine deficiency, iodine toxicity can cause enlargement of the thyroid gland and decreased thyroid hormone synthesis an Toxigoiter can occur.

8.2.3 Zinc

Zinc is primarily intracellular substance. Its total quantity in the body is 2.3g. Largest stores of Zinc is present in the bones. Zinc forms a constituent of the blood. Zinc is an important element performing a range of function in the body as it is a cofactor for a number of enzymes.

Absorption, Transport, Storage, and Excretion of Zinc

☆ Zinc is absorbed throughout the small intestine by simple diffusion and active transport.

☆ When zinc is absorbed into intestinal cells, it induces the synthesis of metallothionein, a protein that binds zinc in much the same way that mucosal ferritin binds iron.

☆ The regulation of zinc absorption may be, in part, related to the synthesis of metallothionein because it hinders the movement of zinc from intestinal cells.

☆ If zinc is not transported out of the intestinal absorptive cell into the bloodstream before the intestinal cells are sloughed off, it passes out of the body in the feces. Thus, a mucosal block, similar to that for iron, decreases excess absorption of zinc. However, large doses of zinc can override the mucosal block. Like iron, the absorption of zinc is affected by diet composition and the body's need for the mineral.

☆ Zinc absorption increases when zinc intake is low or marginal, when animal protein intake is high, and when body needs for zinc are elevated. In contrast, zinc absorption decreases when zinc or non-heme iron in-take is excessive, dietary ber and phytic acid intake is high, and zinc status is adequate.

☆ Zinc absorbed into the bloodstream binds to blood proteins, such as albumin, for transport to the liver. The liver repackages and releases zinc into the blood bound to alpha-2-macroglobulin, albumin, and other proteins.

☆ Although there is no storage site for zinc, the body maintains an exchangeable pool of zinc in the liver, bone, pancreas, kidney, and blood. This allows the body to recycle zinc and maintain zinc status when intake is low.

☆ Excess zinc (unlike iron) is readily excreted through the feces, thus decreasing the risk of toxicity. Small amounts also are excreted in urine and sweat.

Table 8.4: Factors that Affect Zinc Absorption

Sl.No.	Factors that Increase Absorption	Factors that Decrease Absorption
1.	Increased need for zinc	Phytic acid and fibers in whole grains
2.	Animal protein intake	Good zinc status
3.	Zinc deciency	High non-heme iron intake
4.	Low to moderate zinc intake	Excessive zinc intake

Functions of Zinc

1. As many as 300 different enzymes in the body require zinc. Zinc is a constituent of enzymes such as carbonic anhydrase, alkaline phosphatase, lactic dehydrogenase.

2. In fact, it is hard to name a body process or body structure that isn't affected either directly or indirectly by zinc.

3. Zinc contributes to DNA and RNA synthesis,

4. It plays a major role in the alcohol metabolism, heme synthesis, bone formation, acid-base balance, immune function, reproduction, growth and development, and the antioxidant defense network (as a part of the Cu/Zn superoxide dismutase [SOD] enzyme).

5. Zinc also may play a role in shortening the duration of common colds.

6. It is a constituent of the hormone insulin

7. In addition, zinc stabilizes the structures of cell membrane proteins, gene transcription fingers (known as "zinc fingers"), and receptor proteins for vitamin A, vitamin D, and thyroid hormone.

Sources of Zinc

☆ Meat, unmilled cereals and legumes are good sources. Fruits and vegetables are poor sources.

Requirements of Zinc

The daily requirement of Zinc as recommended by the ICMR expert group is given in Table 8.5.

Table 8.3: Requirement of Zinc

Group		Body Weight (kg)	Zinc requirement (mg/day)
Infants			
	0-6 months	5.4	—
	6-12 months	8.4	—
Children	1 – 3 year	12.9	05
	4 – 6 year	18.0	07
	7 – 9 year	25.1	08
Boys and Girls			
	10 – 12 year	34.3 and 35.0	09
	13 – 15 year	47.6 and 46.6	11
	16 – 17 year	55.4 and 52.1	12
Man	60	12	
Woman	55	10	
Pregnancy and Lactation		55	12

Deficiency of Zinc

☆ The symptoms include loss of appetite, delayed growth and sexual maturation, dermatitis, impaired vitamin A function, alopecia, decreased taste sensitivity, poor wound healing, immune dysfunction, severe diarrhea, birth defects, and increased infant mortality.

☆ Compromised zinc status also impairs the integrity of zinc- containing structural proteins in cell membranes, zinc fingers, and protein receptors. As a result, these proteins can no longer perform their functions.

Toxicity of Zinc

Signs of zinc toxicity have been reported with supplemental intakes of zinc at 5 or more times the RDA. Thus, the Upper Level is set at 40 mg/day. The symptoms include loss of appetite, nausea, vomiting, intestinal cramps, and diarrhea. Toxicities also have been reported to impair immune function and reduce copper absorption and the activity of copper-containing enzymes. Individuals taking zinc supplements and/or zinc lozenges (for the relief of cold symptoms) should do so cautiously and with the guidance

8.2.4 Flourine

☆

Fluorine (F_2) is a poisonous gas. The fluoride ion (F^-) is the form of this trace mineral essential for human health. Fluoride, the ionic form of fluorine, may not be an essential nutrient because all basic body functions can occur without it.

☆ However, in the early 1930s, it was observed that individuals living in the south western U.S., where the water naturally contained high concentrations of fluoride, many people in these areas had small spots on the teeth, called mottling due to flouride deposits, these mottled teeth contained very few dental caries (cavities). It was later observed that fluoride in th warwe did indeed decreases the rate of dental caries, so controlled fluoridation of water in many parts of the world is followed.

☆ Flourine is considered as **Double Sword Mineral.** Water should contain the recommended amount of 1ppm or 1mg/L of fluoride. As less amount in body will cause dental caries and excess will cause mottling and fluorosis. Mottling, is referred as discolouration or marking of the surfaces of teeth from exposure to excessive amounts of fluoride (fluorosis).

Absorption, Transport, Storage, and Excretion of Fluoride

☆ The absorption of dietary fluoride occurs rapidly in the stomach and small intestine via passive diffusion.

☆ Overall, approximately 80 to 90 per cent of fluoride consumed is absorbed.

☆ Absorbed fluoride is transported in the bloodstream and concentrated in teeth and the skeleton.

☆ The amount of fluoride deposited in the teeth and bones is greatest during infancy, childhood, and adolescence.

☆ Calcified tissue deposition and urinary excretion are the major means for removing fluoride from the circulation.

Sources of Fluoride

☆ Today, the major source of fluoride is fluoridated water. Typically, fluoridated water contains about 0.7 to 1.2 mg/liter (US FDA).

☆ However, in India not all public or private water sources are fluoridated.

☆ In addition to fluoridated water, tea, seafood, and seaweed provide the greatest amounts of dietary fluoride.

☆ The use of fluoridated toothpastes and mouth rinses and fluoride treatments provides non-dietary ways of obtaining fluoride.

Dietary Needs for Fluoride ESADDI

☆ As per US FDA, the adequate intakes for fluoride are 3 mg/day for adult women and 4 mg/day for adult men.

☆ For infants up to 6 months of age, the adequate Intake is 0.01 mg/day. This

0.7 to 3 mg/day for young children and adolescents. The Adequate Intake recommendations are based on the amount needed to provide resistance to dental caries without causing mottling of the tooth enamel.

Functions of Fluoride

1. Although a truly essential function for fluoride has not been described, fluoride is recognized for its beneficial role in supporting the deposition of calcium and phosphorus in teeth and bones and in protecting against the development of dental caries.

2. Fluoride works in several ways to prevent caries. A dietary intake of fluoride during the development of teeth and bones aids in the synthesis of hydroxy fluorapatite crystals rather than typical hydroxyapatite crystals. These crystals provide greater resistance (than typical hydroxyapatite crystals) to bacteria and acids in the mouth that can erode tooth enamel.

3. Fluoride in the blood contributes to fluoride in the saliva, which promotes the remineralization of enamel lesions and reduces the net loss of minerals from tooth enamel.

Fluoride Deficiency and Toxicity

1. There is no specific deficiency disorder or disease because of insufficient fluoride intake. A pea size amount of fluoridated tooth paste is sufficient to prevent the deficiency.

2. However, a lack of fluoride is associated with an increased incidence of dental caries.

3. Mottling, or fluorosis, of the enamel is the result of chronic intake of excess fluoride during tooth development. Dental fluorosis is not associated with any health risk but does result in discoloration and possible pitting of the enamel.

4. In contrast, fluoride toxicity has been reported in young children who swallow fluoride tablets or solutions. Although rare, acute toxicity can occur rapidly and be life threatening. Thus, fluoridated toothpastes, mouth rinses, and supplements should be kept out of the reach of children to prevent intake of one.

5. The signs of toxicity include nausea, vomiting, diarrhea, sweating, spasms, convulsions, and coma.

6. To minimize the risk of fluorosis, an Upper Level has been set at 0.1 mg/kg body weight/day (0.7 to 2.2 mg/day) for infants and children up to 8 years of age. The Upper Level for children over age 8 and for adults is 10 mg/day.

8.2.4 Copper

The use of copper to treat disease dates back to around 400 B.C. However, the essentiality of copper was not fully recognized until 1964, when evidences of human copper deficiency was reported. Copper has vital functions as a part of many important proteins and enzymes in the body.

Copper in Foods
☆ Copper is found in a variety of foods.

☆ Good sources of copper include liver, shellfish, nuts, seeds, mushrooms, soy products, and dark chocolate. Nuts and legumes are rich sources of copper.

☆ Legumes, whole grain products, and the tap water are important sources of copper.

☆ Meat is only a trivial source of copper, but it may promote copper absorption from other foods, as it does for iron.

Absorption, Transport, Storage, and Excretion of Copper
☆ Copper absorption occurs primarily in the small intestine by simple diffusion (like zinc) and active transport into intestinal absorptive cells and then transported out of the mucosal cells into the bloodstream. Copper absorption is the primary means of regulating copper balance. Thus, absorption can vary from approximately 12 to 70 per cent of dietary intake.

☆ Copper absorption is known to increase when dietary copper is low and to decrease when intakes of copper, iron, and/or zinc are excessive.

☆ In the blood, copper is bound to albumin and other proteins and moves rapidly to the liver (the main storage site) and kidneys.

☆ Copper is transported from the liver to other tissues bound primarily to the protein ceruloplasmin. Within the tissues, ceruloplasmin binds to specific receptors, which release copper to transporters within the cells.

☆ Very little copper is stored in the body. However, excess copper can bind to intestinal metallothionein, which may increase the interim availability of copper.

☆ Copper is excreted mainly through the bile into the GI tract for fecal elimination.

Dietary Requirement for Copper
☆ The adult RDA for copper is 900 µg/day (U.S. RDA).

☆ This allowance is based on the amount needed for the normal activity of copper-containing enzymes and proteins in the body.

Functions of Copper
1. Copper is an important component of many enzymes. Copper-containing enzymes have many functions in metabolism. For example, ceruloplasmin enzyme (also called ferroxidase I) is involved in oxidizing ferrous (Fe^{2+}) iron to ferric (Fe^{3+}) iron for incorporation into transferrin and subsequent transport from the liver to body cells. One effect of low ceruloplasmin levels is that little iron is transported from storage, resulting in decreased hemoglobin synthesis and the development of anaemia. Thus, copper and iron metabolism are closely linked.

2. In combination with zinc, copper also functions as a part of a family of enzymes known as superoxide dismutase (SOD) enzymes. These enzymes eliminate superoxide free radicals, which prevents oxidative damage to cell membranes. Another copper containing enzyme, cytochrome C oxidase, catalyzes the last step of the electron transport chain in energy metabolism.

3. Copper is involved in the regulation of neurotransmitters (serotonin, tyrosine, dopamine, and norepinephrine) by the enzyme monoamine oxidase.

4. Copper also has an important role in connective tissue formation as a component of lysyl oxidase. Lysyl oxidase cross-links the strands in 2 structural proteins (elastin and collagen) that give tensile strength to connective tissues in the lungs, blood vessels, skin, teeth, and bones.

Copper Deficiency

☆ Severe copper deficiency is relatively rare in humans.

☆ Deficiencies have been reported in premature infants fed milk-based formulas, in infants recovering from malnutrition, in patients on long-term total parenteral nutrition without added copper, and in individuals with the genetic disorder Menkes disease. The symptoms of copper deficiency include anaemia, decreased white blood cell counts (leukopenia and neutropenia), and skeletal abnormalities (osteopenia).

☆ Recent studies suggest that copper deficiency may increase the risk of neurological disorders, such as amyotrophic lateral sclerosis (Lou Gehrig's disease) and Alzheimer's disease.

Copper Toxicity

☆ Although copper toxicity is not common in humans, it has been reported in children taking accidental overdoses, in individuals consuming copper-contaminated food or water, and in Wilson's disease (a genetic disorder resulting in excess copper storage).

☆ The symptoms of toxicity include abdominal pain, nausea, vomiting, and diarrhea.

☆ In severe cases, an accumulation of copper in the liver and the brain causes cirrhosis and neurological damage, respectively.

☆ The Upper Level for copper is set at 10 mg/day because higher intakes increase the risk of liver damage.

Questions

Q.1. What functions of minerals in the body?

Q.2. What are the excellent sources of zinc?

Q.3. What are the deficiency diseases of Iron, Zinc and Flourine

Q.4. Why are heme-containing foods more efficient means of obtaining dietary iron than non-heme-containing foods?

Q.5. What factors increase and decrease iron and zinc absorption respectively?

Q.6. Why are girls and women more prone to iron deficiency?

References

1. Wardlaw's, Perspectives in Nutrition, 8th Edition, McGraw-Hill Companies, ISBN 978–0–07–296999–3.

2. Bamji S. Mehtab *et al.,* Textbook of Human Nutrition, 3rd edition, 2010, Oxford and IBH Publishing Co. Pvt. Ltd., New Delhi. ISBN-978-81-204-1742-7.

3. Gopalan.C, B.V. Rama Sastic and Balsosubramaniam. S.C.,2012, reprinted Nutritive Value of Indian Foods. NIN, Hyderabad.

4. Shakuntala Manay, M. and Shadaksharaswamy, M., (1987), Foods-Facts and Principles, New Age International (P) Publishers Ltd., Chennai.

5. Textbook on Nutrition and Dietetics for Higher Secondary, Text Book Corporation, College Road, Chennai. Government of Tamil Nadu, First Edition – 2004.

6. 2010 revised Recommended Dietary Allowances suggested by Indian Council of Medical Research.

7. www.eatright.org

8. www.micronutrient.org/home.asp

9. www.google.com

Unit III

9

Effect of Cooking on Food

Cooking involve subjecting food to heat and when food is subjected to heat then the some chemical changes occur in food and to the nutrients as well. Some changes are beneficial and others are not. Beneficial changes include destruction of microbes, improvement in texture etc. Cooking can also lead to loss of nutritive values but as it improves the shelf life of food so cooking and processing of food is good. Cooking also leads to changes in nutrients also. The chapter outlines the effect of cooking on food:

- ☆ Cooking increases the palatability of food. It make the food more palatable and edible for example raw sweet potato is inedible but cooking (boiling or roasting) makes it edible and palatable.

- ☆ Some basic staple foods such as dry legumes and whole grains are not in edible form when harvested and these products must be rehydrated to soften the texture.

Although meat, fish and poultry are eaten raw in some population but cooking improves the aesthetical appeal and increases the palatability. Also cooking improves the hygiene quality of food and heat destroys the microorganisms. However the extent of microbe destruction depends upon the time and temperature relationship.

- ☆ Digestibility and nutritive value in some cases is increased by cooking. Starch and cooked grain products and legumes become more readily available to digestive enzyme than in compact raw starch granules.

- ☆ Some anti nutrient factors like trypsin inhibitors, saponins, haematoglutenins etc. are present in pulses and legumes. Soaking, germination and cooking removes the antinutritive factors.

☆ Apart from removing antinutritive factors cooking also bring about some decrease in nutritive value as well. Heat sensitive water soluble vitamins (B-vitamins and vitamin-C) get destroyed during cooking.

☆ Cooking and processing of food improves the keeping quality of food as cooking above 70 degee C will kill most of the microbes however pathogenic micro-organism requires a little high temperature. Also, cooked food should be maintained away from temperature danger zone (7-65 degree C) to inhibit the growth of micro-organism.

☆ Cooking also makes it possible show creativity and makes many dishes from same raw material. Thus it adds variety to the diet. For example, rice can be used in making kheer, pulav, boiled rice, biryani, etc.

☆ Processing food also affect flavor, colour and texture of food. New flavours and improvement of texture is observed during baking, roasting, frying etc. for example new flavours are developed during making of bread, caramel and cooking meat. However at the same time some flavours may be lost or undesirable flavours may develop.

☆ Texture of vegetable become soften as fibers of vegetables and connective tissues of meat are tenderized.

☆ Colour changes also occur during cooking. Bright green vegetable turns dull when food is cooked in presence of acid/alkali whereas short blanching actually enhances the colour of fresh green peas. In case of baked product light brown colour and crust formation occurs which improves the appeal of the food.

Questions

Q.1. How does cooking bring changes in the texture of food?

Q.2. What is the effect of cooking on nutrient content of any food?

10
Nutritional Improvement of Food

10.0 Introduction

Food processing makes the food palatable and improves texture of food; at the same time it leads to loss of some nutrients. So this calls for either restoration of lost nutrients or improvement of the nutrient content in the food. This can be achieved by following methods:

- ☆ Fortification
- ☆ Enrichment
- ☆ Supplementation
- ☆ Complementation
- ☆ Gene therapy
- ☆ Parboiling
- ☆ Germination
- ☆ Fermentation

Individual methods are discussed below:

10.1 Fortification

Fortification refers to restoration of lost nutrients due to processing. As per the definition given by WHO/FAO in year 1994. Food fortification refers to the addition of more than one essential nutrient to a food whether or not that nutrient is normally contained in the food, for the purpose of preventing or correcting a demonstrated deficiency of one or more nutrients in the population or specific population groups.

Food fortification is required because of rapidly changing life styles and increasing dependence on more highly processed foods. So, fortification refers to a process of adding something to a food which was not originally present or not present in large quantities. For example this include addition of commercial amino acid to a product where the particular amino acid is absent.

Objective of food fortification are:

1. It restores the nutrient which have been removed during the milling process (bran removal during milling removes the thiamine and other nutrients present in the bran level) or during food processing (Pressing fruits for juice extraction removes the fibre also).
2. To improve nutrient intake levels of target population which are at risk of micronutrient deficiency. In India a micronutrient initiative running to deliver micronutrient to target population.

For example iodized salt, golden rice. Iodization of salt has been taken as a national initiative to correct the deficiency of iodine from the Indian population. This has enable to correct iodine deficiency disorder (IDD) to a certain level in India.

Steps for fortification in developing countries:

1. Identification of requirement of nutritional intervention.
2. Decision on required level of fortification.
3. Selection of suitable carrier (medium and mode)
4. Selection of appropriate fortificant. (form of the nutrient/salt etc.)
5. Determination of the technologies for the fortification process and process optimization.

10.2 Food Enrichment

Although the word Enrichment is used interchangeably with fortification but it is defined as the "restoration of vitamins and minerals lost during processing stage". This process adds nutritional value to meet a specific standard. For example vitamin and mineral enriched bread. This is the process which involves adding back the lost nutrients to the refined product like thiamine, niacin, riboflavin, iron etc.

Refinement is a process that extracts the starch or endosperm of grains, discarding the other cereal parts. Refinement doesn't change the characteristics of particular food. For example Wheat grain is a good source of fiber and nutrients but after refinement of wheat it will mainly a source of calories. So, this increases the need of adding the lost nutrients And the process of adding back the nutrients to the product that were originally present in the grain but were remove during processing is called enrichment.

10.3 Supplementation

Supplementation is the process of adding complete proteins to partially complete/incomplete proteins. Supplementation of limiting essential amino acid with protein concentrates high in amino acids can be done. For ex. cereals are low in lysine. This also includes replacement of the normal cereal grain with its high lysine mutant counterpart. Three high lysine cereals are not available corn, barley and sorghum. *e.g.* Kheer, corn flakes with milk, bread-omellete, porridge with milk etc.

10.4 Complementation

In complementation process two partially complete proteins are given together. Generally plant proteins are mixed with animal proteins which are rich in essential amino acids.

The quality of dietary proteins depends on the pattern of essential amino acids (EAA) it supplies. The best quality protein is the one that has an essential amino acid pattern very close to the pattern of the tissue proteins. Egg proteins and human milk proteins meets these guidelines and as classified as high quality proteins and serve as reference protein for defining the quality of proteins. The proteins of animal foods (milk, meat, fish etc.) are comparable with egg in their essential amino acid composition so, are considered as good quality proteins and have more digestibility. On the other hand, plant proteins are of poorer quality as compared for egg protein as EAA composition is not well balanced and EAA are not present in optimal level present in egg. For example, in comparison with egg protein, cereal proteins are poor in amino acid lysine and pulses and oilseed proteins are rich in lysine but are poor in sulphur containing amino acids. Individually such proteins are called as incomplete proteins. However, this deficiency of particular amino acid of any vegetable food can be conquered by judious blend with other vegetable foods which may have adequate level of that limiting amino acid. Thus amino acid composition of these proteins will complement each other and the resulting mix will have an amino acid pattern better than either of the constituent proteins of the mixture. This practice is usually followed to improve vegetgable proteins quality.

Thus, a deficiency of an amino acid in one can be made-up to an adequate level in another, if both are consumed together. A protein of cereals deficient in lysine, content have a mutually supplementary effect, a combination of cereal and pulse in the ratio of 5:1 has been found to give an optimum combination. For example Balahar which is a combination of cereal and pulses, this is provided in aaganvadi and schools in mid day meal. Other examples are khichdi, idli, dosa, dal-rice.

10.5 Gene Therapy

Gene therapy is the process of improvement of protein quality and quantity by genetic modification.

10.6 Parboiling

Parboiling refers to the process of partial boiling of food as the initial stage in the cooking process. The word is often used in reference to parboiled rice. Parboiling or partial boiling can also be used for removing poisonous or foul-tasting substances from foodstuffs. The technique may also be used to soften vegetables before roasting them. It is also used as a method of dehusking Raw rice or paddy by steam. This steam also partially boils the rice while dehusking. This process generally changes the white colour of rice from to a bit reddish. Because of parboiling the water soluble nutrients (vitamins and minerals) from bran or germ penetrate to endosperm.

Advantages of Parboiling

1. The process imparts a hard texture and a smooth surface finish to the grain as a result which the brokens in the milled rice is minimized. While 90 per cent of the parboiled grains may remain unbroken; the brokens in raw rice could be as high as 50 per cent. The reduction in broken rice results in an increase of 3-5 per cent in the total yield of rice.
2. Insects find it more difficult to gnaw the hard and smooth surface of parboiled rice.
3. The loss of solids in the gruel during cooking is also less in parboiled than in raw rice.
4. The cooking quality is different from that of raw rice. Parboiled rice is non-sticky and non-glutinous.
5. Milled parboiled rice contains more of B-vitamins than milled raw rice. Also, compared to raw rice, loss of B-vitamins is less in parboiled rice during washing and cooking.
6. The parboiled paddy on milling produces a bran higher in oil content (about 25-30 per cent oil) compared to raw rice bran (about 10-20 per cent oil).
8. Parboiled rice bran is relatively stable compared to raw rice.

10.7 Germination

Germination is the process in which plants grows from a seed to new fruit or a flower. The most common example of germination is the sprouting of a seedling from a seed.

During the sprouting process, as the tiny shoot grows from grain, the complex carbohydrates in seeds break down into simple sugars, and the proteins break down into amino acids; and both these primary products are easier for the body to process and absorb. Sprouted foods also cook more quickly. Example of germinated food includes sprouted rice, pulses, lentils etc. As the seeds germinate it builds up or increases vitamins (usually B and C) and other nutrients, often in concentrated form. For example a broccoli sprouts instead of simple broccoli can have up to 50 times more of the cancer-preventing phytochemical sulforaphane. Also, raw sprouts have lots of enzymes.

Advantages of Germinated Food

1. Sprouted rice reduces attack of common allergens and reduces the risk of cardiovascular, sprouted brown rice fights diabetes.
2. Sprouting rye increases and protects folate.
3. Sprouted buckwheat protects against fatty liver and its extract decreases blood pressure.

10.8 Fermentation

Fermentation is a metabolic process in which a micro-organism converts carbohydrate present in food, such as starch or a sugar, into an alcohol or an acid. For example, yeast performs fermentation to obtain energy by converting sugar into alcohol. Bacteria perform fermentation, converting carbohydrates into lactic acid. This process has been used to produce bread, wine, cheese, beer, yogurt and other fermented products like sauerkraut, kimchi and pepperoni.

Advantages of Fermentation

Fermentation process is used to preserves the food, increase the shelf life of product, nutrients, improve digestion and creates beneficial enzymes, produce B and C-vitamins, increases the flavor, Omega-3 Fatty acids and various strains of probiotics. Apart from the nutrition aspect the fermented products are economical.

10.9 Conservation of Nutrients in Preparation and Cooking

Loss of nutrients in food (vegetable, cereals, pulses etc.) begins from preparation onward and is greater during the cooking process.

1. Peeling of fruits and vegetables leads to loss of vitamins present under the skin.
2. Nutrients are also lost when the edible leaves of carrot, beetroot and outer layer of cabbage, stems of leafy vegetables are discarded.
3. Vitamin B complex and Vitamin C are water soluble vitamins and are lost when the water in which vegetables are cooked and water in which pulses are soaked is discarded. Sodium, potassium and chlorine are also lost when cooking water is discarded.
4. Addition of soda results in heavy loss of B-Vitamins during cooking.
5. Vitamin C is lost by oxidation due to exposure of air to cut fruits and vegetables.
6. During drying ascorbic acid and carotene are lost.

How to Minimize Nutrient Losses during Preparation

1. Wash vegetables before cutting. Soaking or washing time should be reduced to minimize nutrient loss.
2. Use a vegetable peeler to remove skin as it helps remove only a very thin layer of skin.
3. Cut vegetables into big pieces so that exposure of vitamins to water is less while cooking and washing.
4. Minimum water should be used for cooking. Bring the water to boil and then add the vegetables to cook.
5. Cover the vessel with a lid while cooking as it hastens cooking.
6. Cook vegetables by steaming and pressure-cooking to preserve nutrients.

7. Soda or sodium bicarbonate should not be used while cooking vegetables as it destroys valuable vitamins for example during boiling rajmah etc.

8. Vegetables salads should be prepared just before serving to conserve nutrients.

9. Add acids such as lime juice or vinegar to salads as it prevents loss of Vitamin C.

References

1.   Dietary guidelines for Indians- A Manual, 2nd Edition, 2011, National Institute of Nutrition, Indian Council of Medical Research, Hyderabad, India.

2.   http://www.foodandwine.com/articles/sprouting-a-germination-nation

3.   chemistry.about.com/od/lecturenoteslab1/f/What-Is-Fermentation.htm

4.   http://chemistry.about.com/od/lecturenoteslab1/f/What-Is-Fermentation.htm

5.   http://wellnessmama.com/2245/health-benefits-of-fermented-foods/

6.   http://www.rkmp.co.in/content/advantages-of-parboiling

7.   http://en.wikipedia.org/wiki/Parboiling

11 Labelling

11.0 Introduction

'Labelling' means any words, particulars, trade marks, brand name, pictorial matter or symbol describing a foodstuff and placed on any packaging, document, notice, label etc. accompanying or referring to such foodstuffs. Label refers to any information or pictorial device written, printed, or graphic matter may be displayed in the label as per the FSS regulation. The information provided on the label must not be false or misleading, must be easy to understand, be clearly legible, prominent, definite, plain, unambiguous, it must also be indelible, readily legible by the consumer under normal conditions of purchase, easy to see and not obscured in any way. Food products, including food imports sold in India must be labelled in English or Hindi (with optional labelling in regional language).

Nutritional labelling refers to the standardized display of nutrient content of the food on the label. There are few broad categories of claim:

1. **Nutrition Claims:** This claim refers to the representation which state, suggest or imply that a food product has a particular nutritional properties including energy value, amount of carbohydrate, fat, protein, vitamins and minerals. Under this it is important to mention that list of ingredients and mandatory declaration as per FSSA of certain nutrients cannot be treated as a nutrient claim.

2. **Nutrient Function Claim:** This claim describes the physiological role of a nutrient in the growth, development, and normal functioning of the body for example calcium (mineral) is required for stronger bones, teeth.

3. **Nutrient Content Claim:** This claim describes the level of a nutrient present or available in a food. The difference in energy value and nutrient content should be given under this claim.

4. Comparative Claim: This claim compares the nutrient levels or energy value of two or more foods. Under this claim, different varieties of the same food or similar food are compared and the difference in nutrient content and energy value should be given.

Nutrition Labelling

All individuals require food; food contains nutrients which help to prevent various diseases and to recover from illness. Consumers must know what they are eating with respect to the nutrition or amount of various nutrients present in that food so that they can make their choice to purchase a particular food thus display of nutritional information on the label will help the consumers to choose healthier diets and reduce the risk of diseases related with deficiency of nutrients in food.

Considering the importance of nutritional labelling, in 1990, the U.S. Congress passes the Nutrition Labelling and Education Act (NLEA). Under this law it was required that FDA should develop requirements for the nutritional labelling for almost all foods sold in grocery shop in the United States. Along with the "Nutritional Facts", U.S. FDA permitted the use of few related terms with nutrient content claim on the food label, this includes terms like "free", "light", "low", and "high" having their specific meaning as given in the Table 11.1.

11.1 Why do we Need Nutritional labelling?

The main aim of any government regulating food and food industry is to ensure that the consumer are provided with complete, required and useful information about the food products they purchase. The information will allow the consumer to compare products in terms of economic and health reason.

1. Nutrition labels can help a customer to decide between products and keep a check on the amount of foods a consumer is eating (that are high in fat, salt and added sugars).
2. Nutrition labels can also help you to choose a more balanced diet.
3. The food industry contributes to educating consumers by informing them about product ingredients, nutritional content and health claims etc.
4. Ingredient labelling on the product will also allow the consumers to avoid foods which contain any ingredient which may pose any health hazard to their health. For example any product which contain any ingredient (like colours, gluten, peanut, etc.) which may create hyperallergic reactions in individuals.

In India, there is a regulatory authority named as Food Safety and Standards Authority of India (FSSAI) which has been established under Food Safety and Standards Act, 2006 (Act 34 of 2006) with a function to consolidate various acts and orders that were applicable to food and food products (before 2006), and to handle various food safety and standards related issues in various Ministries and Departments which were handling earlier acts and orders leading to overlapping and misunderstanding among the food industry/food handler. FSSAI has been

Table 11.1: Terms Used for Nutritional Labelling

I. Related with Sugar

1. Sugar free — Less than 0.5 grams per serving

2. No added sugar, without added sugar, no sugar added —
   - ☆ No sugars added during processing or packing, including ingredients that contain sugars (for example, fruit juices, dried fruit or apple sauce).
   - ☆ The food that it resembles and for which it substitutes normally contains added sugar.
   - ☆ Processing does not increase the sugar content above the amount naturally present in the ingredients.
   - ☆ If the food doesn't meet the requirements for a low-or reduced calorie food, the product bears a statement that the food is not low calorie or calorie reduced.

3. Reduced Sugar — Atleast 25 per cent less sugar per serving than reference food.

II. Related with Calories

1. Calorie free — Fewer than 5 calories per serving

2. Low Calorie — 40 calories or less per serving and if the serving is 30g or les or 2 tablespoons or less, per 50g of the food.

3. Reduced or Fewer Calories — Atleast 25 per cent fewer calories per serving than reference food

III. Related with Fat

1. Fat Free — Less than 0.5g of fat per serving.

2. Saturated Fat free — Less than 0.5g per serving and the level of trans fatty acids doesn't exceed 1 per cent of total fat.

3. Low Fat — 3g or less per serving, and if the serving is 30g or less or 2 tablespoons or less per 50g of the food.

4. Low Saturated Fat — 1g or less per serving and not more than 15 percent of calories from saturated fatty acids.

5. Reduced or Less Fat — Atleast 25 percent less per serving than reference food.

6. Reduced or Less Saturated Fat — Atleast 25 percent less per serving than reference food.

IV. Related with Cholesterol

1. Cholesterol Free — Less than 2milligram (mg) of cholesterol and 2g or less of saturated fat per serving.

2. Low Cholesterol — 20mg or less and 2g or less of saturated fat per serving and it the serving is 30g or less or 2 tablespoons or less, per 50g of the food.

3. Reduce or Less Cholesterol — Atleast 25 percent less and 2 g or less of saturate fat per serving than reference food.

Contd...

Table 11.1 –*Contd...*

V. Related with Fiber

1.	High Fiber	5g or more per serving. (Foods claiming high fiber claims must meet the requirement for low fat, or the level of total fat must appear next to the high fiber claim.
2.	Good Source of Fiber	Foods which gives 2.5g to 4.9g fiber per serving
3.	More or Added fiber	Atleast 2.5g more per serving than reference food.

VI. Related with Sodium

1.	Sodium Free	Foods which give less than 5mg per serving.
2.	Low Sodium	140 mg or less per serving and, if the serving is 30g or less or 2 tablespoons or less, per 50g of the food.
3.	Very Low Sodium	35mg or less per serving and if the serving is 30g or less or 2 tablespoons or less, per 50g of the food.
4.	Reduced or Less Sodium	Atleast 25 percent less per serving than reference food.

VII. Related with 'gluten-free' for food labeling

1.	**'Gluten-free'**	The FDA was directed to issue the new regulation by the Food Allergen Labeling and Consumer Protection Act (FALCPA), which directed FDA to set guidelines for the use of the term gluten-free to help people with celiac disease. The U.S. Food and Drug Administration (FDA) recently in the year 2013 published a new regulation in the *Federal Register* defining the term "gluten-free" for voluntary food labelling. This new federal definition standardizes the meaning of gluten-free claims across the food industry. It requires that, in order to use the term gluten-free on its label, a food must meet all of the requirements of the definition, including that the **food must contain less than 20 parts per million of gluten.** Food manufacturers will have a year after this rule is published to bring their labels into compliance with the new requirements to help the people with celiac disease to maintain a gluten-free diet.

Source: Food Science, 5th edition, by Norman N. Potter, Joseph H. Hotchkiss.

created for laying down science based standards for articles of food and to regulate their manufacture, storage, distribution, sale and import to ensure availability of safe and wholesome food for human consumption. FSSAI has proposed regulations on packaging and labelling named as FOOD SAFETY AND STANDARDS (PACKAGING AND LABELLING) REGULATIONS' 2011, which came in force from 05.August.2011 and since then guides the food handler or manufacturer on packaging labelling. Food Safety and Standards Authority of India is under Ministry of Health and Family Welfare.

A label carry mandatory information as required by the food safety and standards (packaging and labelling) regulations' 2011 which includes Mandatory and some Optional Information as mentioned below :

Mandatory	*Optional/Voluntary*
Name of food, Category of Food, Ingredient List, Nutritional Labeling, Consumer Contact details, Instruction for Use, MRP, Lot No., Net Weight, Best before and MFD date, Addresses - MKTD/MFG/PKD By, Storage Conditions. For Imported Product - Country of origin and Imported By.	Bar Code, Allergen Information, Recycle Symbol etc.

Changes have been observed in the international trade of food and other agricultural commodities which includes :

1. Increase in international trade of both raw agricultural commodities and processed foods.
2. Decrease in traditional economic trade restrictions and tariffs that are imposed on foods and agricultural products and other goods.
3. Increase in the FDI.

These changes have lead to the requirement of setting international standards to ensure that the products meet country's safety, quality and product standards. World Health Organization (WHO) and Food and Agricultural Organization (FAO) of the United Nations are the two major agencies who are working towards increasing and improving food resources, nutrition and health throughout the world. The need for coordination in setting standards was recognised which in 1962 lead the establishment of Codex Alimentarius Commission (CAC) under the auspices of the United Nations through FAO/WHO.

Thus, the global level, Codex Alimentarius Commission develops agreements on international standards and safety measure for foods and agricultural products. CAC propound international food standards, guidelines and codes of practice contributing to the safety, quality and fairness of international food trade. Codex sets minimum quality, safety, and hygienic standards which is of voluntary nature and the countries voluntarily adhere to the standards for importing and exporting food products. Codex adopted guidelines to nutritional labeling in 1985. It says that label is to inform consumers about the nutritional content of foods, act as information conveying medium. India is in the process of harmonizing standards with codex alimentarius.

In India before the implementation of food safety and standards (packaging and labelling) regulations' 2011, following acts were applicable which were ruling food and food industry:

1. **Prevention of Food Adulteration (PFA) Act'1954** rule 32 and 42 was regulating the labelling requirement of various foods.

2. **Edible Oil Packaging Order'1988** defined that the edible oils should be packed in a container, marked and labelled in manner specified in schedule A of the Act.

3. **Milk and Milk Products Order' 1992 (MMPO)** laid the packing and labelling requirements for which the milk and milk product producers should comply.

4. **Fruit Products Order'1955 (FPO)** was to be complied for packing and labelling of fruit products.

5. **Meat Food Product Order'1973 (MFPO)** regulated the production, quality and distribution of raw and processed meat food products and also regulated the labelling guidelines. For example every package containing monosodium glutamate (MSG) shall bear the label saying "This package contains MSG and is unfit for infants below 12 months". Label was also required to carry the licence number and category.

However due to complexity and overlapping of various acts and orders, and lacunas like existing acts were not focussed on food safety of street foods and traceability (to a very less extent). So, the need arose to have one act/order to govern food safety and standards related issues, that is how the FSSAI came into action and the authority is working very hard to tackle the issues and ensure that consumers gets a safe food.

11.2 Where should a Label be Placed on a Package?

1. Nutrition labels on the back or side of packaging

   In this pattern the nutritional information is given on back or side of the product pack. For example, the image below shows the back of pack nutrition label on a loaf of white bread and chocolate box.

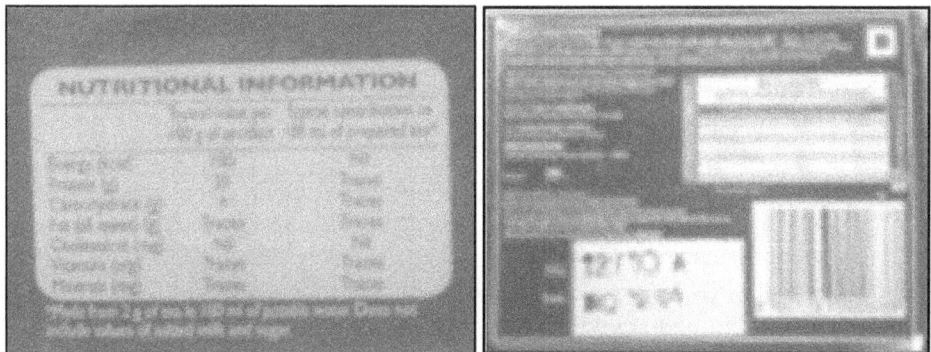

Figure 11.1: Nutritional Information on Back Panel.

2. This type of label usually includes information on energy (kJ/kcal), protein, carbohydrate and fat. It may also provide additional information on saturated fat, sugars, sodium, salt and fibre. All nutrition information is provided per 100 grams and sometimes per portion of the food. Figure 11.1 shows Nutrition labels on the back or side of packaging.

3. Nutrition labels on the front of packaging

   Many food manufacturers/processor, big supermarkets also exhibit nutritional information on the front of pre-packed food. This is very useful for the customer, as this will help them to compare different food products at a glance.

   Labels with information given on front of pack (such as in above image of the label of honey) usually provide information in a serving or portion of the food on energy (in kJ and kcal), fat (in grams), saturated fat (in grams), sugar (in grams) and salt (in grams) content. Some of front pack nutrition labels also provide information about RI. Table 11.2 shows a nutritional label on front of packaging.

Table 11.2: Nutritional Information on Front Panel
Each 160g portion provides

Energy 672 kJ 160kcal	Fat 4.6g	Saturated Fat 2.6g	Salt 0.16g	Sugars 18.4g
8 per cent	7 per cent	13 per cent	3 per cent	20 per cent

Of reference intake Typical values per 100g: Energy 420kJ/100kcal.

11.3 Labeling Requirements

11.3.1. Following are some important **definition** to enable the interpreter to understand the labelling requirements of the regulation.

1. **"Date of manufacture"** means the date on which the food becomes the product as described; The manufacturer/producer can display the date in any format like DD/MM/YEAR, DD.MM.YEAR, YEAR.MM.DD etc.

2. **"Best before"** means the date which signifies the end of the period under any stated storage conditions during which the food shall remain fully marketable and shall retain any specific qualities for which tacit or express claims have been made and beyond that date, the food may still be perfectly safe to consume, though its quality may have diminished. However the food shall not be sold if at any stage the product becomes unsafe.

   The manufacturer/producer can display the date in any format like DD/MM/YEAR, DD.MM.YEAR, YEAR.MM.DD etc. The manufacturer can also mention the time of best before.

3. **"Date of packaging"** means the date on which the food is placed in the immediate container in which it will be ultimately sold;

For example during festival season various companies launch their products for gifting purpose in nice attractive packs which contains individual units of their product. In that case a manufacturing date is given on the individual pack or units and date of packaging would be given on the outer pack.

4. **"Infant"** means a child not more than twelve months of age.

5. **"Lot number" or "code number" or "batch number"** means the number either in numericals or alphabets or in combination thereof, representing the lot number or code number or batch number, being preceded by the words "Lot No" or "Lot" or "code number" or "Code" or Batch No" or "Batch" or any distinguishing prefix by which the food can be traced in manufacture and identified in distribution.

Apart from manufacturing date, Lot Number may also carry the information pertaining to the place of manufacturing (in case if the company has multiple manufacturing sites in the country) machine number, production line and shift of manufacturing and even about machine operator. An example of batch code the label shown in picture no 3 is B1CF G50.

6. **"Multipiece package"** means a package containing two or more individually packaged or labelled pieces of the same commodity of identical quantity, intended for retail either in individual pieces or packages as a whole.

For example chocolate gift packs, biscuit small units sold in a bigger pack.

7. **"Non-vegetarian Food"** means an article of food which contains whole or part of any animal including birds, fresh water or marine animals or eggs or products of any animal origin, but excluding milk or milk products, as an ingredient;

8. **"Prepackaged" or "Pre-packed food"**, means food, which is placed in a package of any nature, in such a manner that the contents cannot be changed without tampering it and which is ready for sale to the consumer.

9. **"Principal Display Panel" (PDP),** means that part of the container/package which is intended or likely to be displayed or presented or shown or examined by the customer under normal and customary conditions of display, sale or purchase of the commodity contained therein. It is normally considered the "front" side of the panel. A general layout of the label showing PDP is shown below:

General Information Panel	Principle Display Panel	Nutritional Information Panel

10. **"Use – by date" or "Recommended last consumption date" or "Expiry date"** means the date which signifies the end of the estimated period under any stated storage conditions, after which the food probably will not have

the quality and safety attributes normally expected by the consumers and the food shall not be sold;

This is normally given for milk and other perishable commodities as after the expiry date the product is not fit for consumption and the quality would not be such as intended by the consumer.

11. **"Vegetarian Food"** means any article of Food other than Non- Vegetarian Food as defined above in point 7.

12. **"Wholesale package"** means a package containing:

(a) a number of retail packages, where such first mentioned package is intended for sale, distribution or delivery to an intermediary and is not intended for sale direct to a single consumer; or

(b) a commodity of food sold to an intermediary in bulk to enable such intermediary to sell, distribute or deliver such commodity of food to the consumer in smaller quantities.

11.3.2 General Requirements

1. Every prepackaged food shall carry a label containing information as required below :

2. The particulars of declaration required under these Regulations to be specified on the label shall be in English or Hindi in Devnagri script.

   This doesn't mean that other languages cannot be used on the label. Instead the regulation permit the manufacturer to use any other language (regional language) along with the language required under this regulation.

3. Pre-packaged food shall not be described or presented on any label or in any labelling manner that is false, misleading or deceptive or is likely to create an erroneous impression regarding its character in any respect.

4. Label in pre-packaged foods shall be applied in such a manner that they will not become separated from the container.

5. Contents on the label shall be clear, prominent, indelible and readily legible by the consumer under normal conditions of purchase and use.

6. Where the container is covered by a wrapper, the wrapper shall carry the necessary information or the label on the container shall be readily legible through the outer wrapper and not obscured by it.

11.3.3 Labelling of Pre-packaged Foods

In addition to the General Labelling requirements specified in 11.3.2 above every package of food shall carry the following information on the label, namely:

1. **The Name of Food:** The name of the food shall include trade name or description of food contained in the package.

2. **List of Ingredients**: Except for single ingredient foods, a list of ingredients shall be declared on the label in the following manner:

(a) The list of ingredients shall contain an appropriate title, such as the term "Ingredients";

(b) The name of Ingredients used in the product shall be listed in descending order of their composition by weight or volume, as the case may be, at the time of its manufacture;

(c) A specific name shall be used for ingredients in the list of Ingredients;

Provided that for Ingredients falling in the respective classes, the following class titles may be used as given in Table 11.3, namely:

Table 11.3: The Class Titles of Food Ingredients

Classes	Class Titles
Edible vegetable oils/Edible vegetable fat	Edible vegetable oil/Edible vegetable fat or both hydrogenated or Partially hydrogenated oil
Animal fat/oil other than milk fat	Give name of the source of fat. Pork fat, lard and beef fat or extracts thereof shall be declared by specific names
Starches, other than chemically modified starches.	Starch
All species of fish where the fish constitutes an ingredient of Fish another food and provided that the labelling and presentation of such food does not refer to a species of fish.	Fish
All types of poultry meat where such meat constitutes an ingredient of another food and provided that the labelling and presentation of such a food does not refer to a specific type of poultry meat.	Poultry meat
All types of cheese where cheese or mixture of cheeses constitutes an ingredient of another food and provided that the labelling and presentation of such food does not refer to a specific type of cheese.	Cheese
All spices and condiments and their extracts.	Spices and condiments or mixed spices/condiments as appropriate
All types of gum or preparations used in the manufacture of gum base for chewing gum.	Gum Base
Anhydrous dextrose and dextrose monohydrate.	Dextrose or Glucose
All types of Caseinates.	Caseinates
Press, expeller or refined cocoa butter.	Cocoa butter
All crystallized fruit.	Crystallized fruit
All milk and milk products derived solely from milk.	Milk solids
Cocoa bean, Coconib, Cocomass, Cocoa press cakes, Cocoa powder (Fine/Dust).	Cocoa solids
Pork fat, lard and beef fat or extract thereof	Shall be declared by their specific names;

(d) **For declaring Compound Ingredients:** Where an ingredient itself is the product of two or more ingredients, such a compound ingredients shall be declared in the list of ingredients, and shall be accompanied by a list in brackets, of its ingredients in descending order of weight or volume, as the case may be. In products where a compound ingredient, constitutes less than 5 per cent of the food, the list of ingredients of the compound ingredient, other than food additive, need not to be declared.

(e) Added water shall be declared in the list of ingredients except in cases where water forms part of an ingredient, such as, brine, syrup or broth, used in the compound food and so declared in the list of ingredients. Water or other volatile ingredients evaporated in the course of manufacture need not be declared.

In the case of dehydrated or condensed foods, which are intended to be reconstituted by addition of water, the ingredients in such reconstituted food shall be declared in descending order of weight or volume as the case may be, and shall contain a statement such as "Ingredients of the product when prepared in accordance with the directions on the label".

(f) Every package of food sold as a mixture or combination shall disclose the percentage of the ingredient used at the time of the manufacture of the food (including compound ingredients or categories of ingredients), if such ingredient:

(i) is emphasised as present on the label through words or pictures or graphics for example as displayed in Figure 11.2.

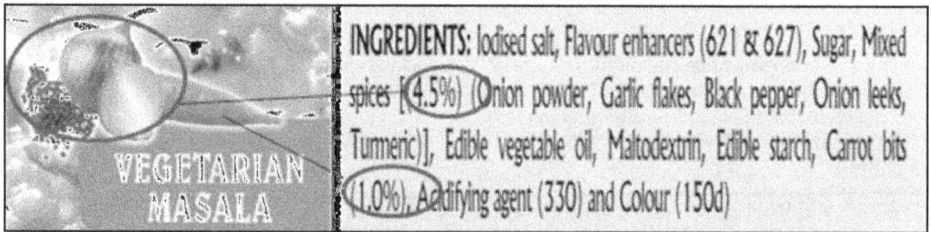

INGREDIENTS: Iodised salt, Flavour enhancers (621 & 627), Sugar, Mixed spices (4.5%) (Onion powder, Garlic flakes, Black pepper, Onion leeks, Turmeric)], Edible vegetable oil, Maltodextrin, Edible starch, Carrot bits (1.0%), Acidifying agent (330) and Colour (150d)

Figure 11.2: Part of Label Showing Picture of Onion and Carrot Bits and Amount of same in the Ingredient List.

(ii) or is not within the name of the food but, is essential to characterise the food and is expected to be present in the food by consumers, and if the omission of the quantitative ingredient declaration will mislead or deceive the consumer.

Where the ingredient has been used as flavouring agent, the disclosure of such ingredient is not required. Further where the drained net weight is indicated on the label as required or in case of such food products where specific provisions are stipulated under these regulations or where a pictorial representation of a serving suggestion is made for consumer information and use, the disclosure of such ingredient is not required.

In case of any bottle containing liquid milk or liquid beverage having milk as an ingredient, soft drink, carbonated water or ready-to-serve fruit beverages, the declarations with regard to addition of fruit pulp and fruit juice shall invariably appear on the body of the bottle.

3. **Nutritional information** – Nutritional Information or nutritional facts per 100 gm or 100ml or per serving of the product shall be given on the label containing the following:

  (i) energy value in kcal;

 (ii) the amounts of protein, carbohydrate (specify quantity of sugar) and fat in gram (g) or ml;

(iii) the amount of any other nutrient for which a nutrition or health claim is made.

When a claim is made regarding the amount or type of fatty acids or the amount of cholesterol, the amount of saturated fatty acids, monounsaturated fatty acids and polyunsaturated fatty acids in gram (g) and cholesterol in milligram (mg) then it shall be declared on the label, and the amount of trans fatty acid in gram (g) shall be declared in addition to the other requirement stipulated above.

(iv) Wherever, numerical information on vitamins and minerals is declared, it shall be expressed in metric units.

 (v) Where the nutrition declaration is made per serving, the amount in gram (g) or milliliter (ml) shall be included for reference beside the serving measure.

Provided that the food claimed to be enriched with nutrients, such as, minerals, proteins, vitamins, metals or their compounds, amino acids or enzymes shall give the quantities of such added nutrients on the label.

Further regulation has provided that

  (i) The nutritional information may not be necessary, in case of foods such as raw agricultural commodities, like, wheat, rice, cereals, spices, spice mixes, herbs, condiments, table salt, sugar, jaggery, or non – nutritive products, like, soluble tea, coffee, soluble coffee, coffee-chicory mixture, packaged drinking water, packaged mineral water, alcoholic beverages or fruit and vegetables, processed and pre-packaged assorted vegetables, fruits, vegetables and products that comprise of single ingredient, pickles, papad, or foods served for immediate consumption such as served in hospitals, hotels or by food services vendors or halwais, or food shipped in bulk which is not for sale in that form to consumers.

 (ii) The compliance to quantity of declared nutrients on the label shall be according to the established practices. For the purpose of this provision, at the time of analysis, due consideration, based on shelf-life, storage, and inherent nature of the food shall be kept in view in case of quantity declared nutrients.

(iii) The food, in which hydrogenated vegetable fats or bakery shortening is used shall declare on the label that '**hydrogenated vegetable fats or bakery shortening used, contains trans fats**.

A health claim of **'trans fat free'** may be made in cases where the trans fat is less than 0.2 gm per serving of food and the claim **'saturated fat free'** may be made in cases where the saturated fat does not exceed 0.1 gm per 100 gm or 100 ml of food.

For the purpose of regulation 11.3.3 (3);

(i) **"Health claims"** means any representation that states, suggests or implies that a relationship exists between a food or a constituent of that food and health and include nutrition claims which describe the physiological role of the nutrient in growth, development and normal functions of the body, other functional claims concerning specific beneficial effect of the consumption of food or its constituents, in the context of the total diet, on normal functions or biological activities of the body and such claims relate to a positive contribution to health or to the improvement of function or to modifying or preserving health, or disease, risk reduction claim relating to the consumption of a food or food constituents, in the context of the total diet, to the reduced risk of developing a disease or health related condition. Figure 11.3 shows a part of label giving health claim.

(ii) **"Nutrition claim"** means any representation which states, suggests or implies that a food has particular nutritional properties which are not limited to the energy value but include protein, fat carbohydrates, vitamins and minerals. Figure 11.3 shows a part of label giving nutritional claim.

(iii) **"Risk reduction"** in the context of health claims means significantly altering a major risk factor for a disease or health-related condition.

Further it is required in the case of returnable new glass bottle manufactured and used for packing of such beverages on or after 19th March 2009, the list of ingredient and nutritional information shall be given on the bottle.

Nutritional claim for added calcium, rich in iron, fortified with vitamins and minerals, nutritionally complete.

Figure 11.3 Part of Label giving Health Claim.

4. Declaration Regarding Veg or Non-veg –

(i) Every package of "Non-vegetarian" food shall bear a declaration to this effect made by a symbol and colour code as stipulated below to indicate that the product is Non-vegetarian Food. The symbol shall consist of a brown colour filled circle having a diameter not less than the minimum size specified in the regulation as mentioned in table no. 11.3, inside a square with brown outline having sides double the diameter of the circle as indicated below :

Brown colour:

(ii) Where any article of food contains egg only as Non-Vegetarian ingredient, the manufacturer, or packer or seller may give declaration to this effect in addition to the said symbol.

(iii) Every package of Vegetarian Food shall bear a declaration to this effect by a symbol and colour code as stipulated below for this purpose to indicate that the product is Vegetarian Food. The symbol shall consist of a green colour filled circle, having a diameter not less than the minimum size specified in the Table 11.4, inside the square with green outline having size double the diameter of the circle, as indicated below:

Green colour:

(iv) Size of the logo

Table 11.4: The Size of Logo

S.No	Area of Principal Display Panel	Minimum Size of Diameters in mm
1.	Upto 100 cms square	3
2.	Above 100 cms square upto 500 cms square	4
3.	Above 500 cms square upto 2500 cms square	6
4.	Above 2500 cms square	8

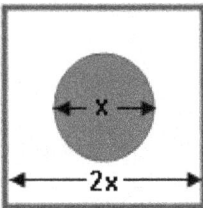

'x : Diameter of inner circle'

The symbol shall be prominently displayed:

(i) on the package having contrast background on principal display panel;

(ii) just close in proximity to the name or brand name of the product;

(iii) on the labels, containers, pamphlets, leaflets, advertisements in any media;

The provisions of regulation 11.3.3 (4) shall not apply in respect of mineral water or packaged drinking water or carbonated water or alcoholic drinks, or liquid milk and milk powders.

5. Declaration regarding Food Additives

(i) For food additives falling in the respective classes and appearing in lists of food additives permitted for use in foods generally, the following class titles shall be used together with the specific names or recognized international numerical identifications:

Acidity Regulator, Acids, Anticaking Agent, Antifoaming Agent, Antioxidant, Bulking Agent, Colour, Colour Retention Agent, Emulsifier, Emulsifying Salt, Firming Agent, Flour Treatment Agent, Flavour Enhancer, Foaming Agent, Gelling Agent, Glazing Agent, Humectant, Preservative, Propellant, Raising Agent, Stabilizer, Sweetener, Thickener.

(ii) Addition of colours and/or Flavours:

(a) Extraneous addition of colouring matter should be mentioned on the label. When an extraneous colouring matter is added to any article of food, then it shall be displayed by one of the following statements in capital letters, just beneath the list of the ingredients on the label attached to any package of food so coloured, namely:

CONTAINS PERMITTED NATURAL COLOUR(S)

OR

CONTAINS PERMITTED SYNTHETIC FOOD COLOUR(S)

OR

CONTAINS PERMITTED NATURAL AND SYNTHETIC FOOD COLOUR(S)

Above statement is displayed along with the name or INS no of the food colour, the colour used in the product need not be mentioned in the list of ingredients.

(b) Extraneous addition of flavouring agents to be mentioned on the label.

Where an extraneous flavouring agent has been added to any article of food, it shall be written just beneath the list of ingredients on the label attached to any package of food so flavoured, a statement in capital letters as below and indicated on a product label Figure 11.4.

CONTAINS ADDED FLAVOUR
(specify type of flavouring agent as per Regulation 3.1.10(1) of
Food Safety and Standards
(Food product standards and food additive) Regulation, 2011

Ingredients: Sugar, Milk solids, Instant coffee-chicory mixture, Maltodextrin, Cocoa solids (1.0%), Caseinate (Milk protein) and Salt.

Ingredients After Reconstitution: Water, Sugar, Milk solids, Instant coffee-chicory mixture, Maltodextrin, Cocoa solids, Caseinate (Milk protein) and Salt.

CONTAINS ADDED NATURE IDENTICAL FLAVOURING SUBSTANCE.

Figure 11.4: Part of a Label sShowing the Declaration of Added Flavour.

(c) In case both colour and flavour are used in the product, one of the following combined statements in capital letters shall be displayed, just beneath the list of ingredients on the label attached to any package of food so coloured and flavoured, as:

CONTAINS PERMITTED NATURAL C0LOUR(S) AND ADDED FLAVOUR(S)

OR

CONTAINS PERMITTED SYNTHETIC FOOD COLOUR(S)
AND ADDED FLAVOUR(S)

OR

CONTAINS PERMITTED NATURAL AND SYNTHETIC FOOD COLOUR(S
AND ADDED FLAVOUR(S)

Above statement is depicted on the product label as given in Figure 11.5

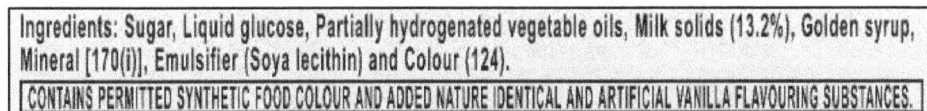

Ingredients: Sugar, Liquid glucose, Partially hydrogenated vegetable oils, Milk solids (13.2%), Golden syrup, Mineral [170(i)], Emulsifier (Soya lecithin) and Colour (124).
CONTAINS PERMITTED SYNTHETIC FOOD COLOUR AND ADDED NATURE IDENTICAL AND ARTIFICIAL VANILLA FLAVOURING SUBSTANCES.

Figure 11.5: Part of a Label Showing the Declaration of Added Colour and Flavour.

In case of artificial flavouring substances, the label shall declare the common name of the flavours, but in case of the natural flavouring substances or nature identical flavouring substances, the class name of flavours shall be mentioned on the label and it shall comply with the requirement of label declaration as specified under 11.3.3 (5)(ii).

Note: When statement regarding addition of colours and/or flavours is displayed on the label in accordance with regulation 11.3.3 (5)(ii) and regulation 3.2.1 of Food Safety and Standards (Food Product Standards and Food Additive) Regulation, 2011, addition of such colours and/or flavours need not be mentioned in the list of ingredients.

Also, in addition to above statement, the common name or class name of the flavour shall also be mentioned on label. Provided further that when combined declaration of colours and flavours are given, the International Numerical Identification number of colours used shall also be indicated either under the list of ingredients or along with the declaration.

Further every package of synthetic food colours preparation and mixture shall bear a label upon which is printed a declaration giving the percentage of total dye content.

6. Name and complete address of the manufacturer

(i) The name and complete address of the manufacturer and the manufacturing unit if these are located at different places and in case the manufacturer is not the packer or bottler, the name and complete address of the packing or bottling unit as the case may be shall be declared on every package of food as shown in the Figure 11.6.

(ii) Where an article of food is manufactured or packed or bottled by a person or a company under the written authority of some other manufacturer or company, under his or its brand name, the label shall carry the name and complete address of the manufacturing or packing or bottling unit as the case may be, and also the name and complete address of the manufacturer or the company, for and on whose behalf it is manufactured or packed or bottled. Example is given in the Figure 11.6.

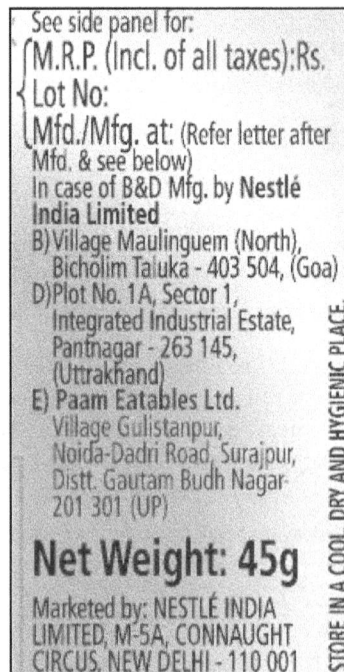

Figure 11.6: Part of a Label Showing the Address.

(iii) Where an article of food is imported into India, the package of food shall also carry the name and complete address of the importer in India. Provided further that where any food article manufactured outside India is packed or bottled in India, the package containing such food article shall also bear on the label, the name of the country of origin of the food article and the name and complete address of the importer and the premises of packing or bottling in India.

7. Net quantity

(i) Net quantity by weight or volume or number, as the case may be, shall be declared on every package of food (refer picture no. 11.7); and

(ii) In addition to the declaration of net quantity, a food packed in a liquid medium shall carry a declaration of the drained weight of the food.

Explanation 1. For the purposes of this requirement the expression "liquid medium" include water, aqueous solutions of sugar and salt, fruit and vegetable juices or vinegar, either singly or in combination.

Explanation 2. In declaring the net quantity of the commodity contained in the package, the weight of the wrappers and packaging materials shall be excluded.

(iii) Where a package contains a large number of small items of confectionery, each of which is separately wrapped and it is not reasonably practicable to exclude from the net weight of the commodity. In such case, the weight of such immediate wrappers of all the items of the confectionery contained in the package, the net weight declared on the package containing such confectionary or on the label thereof may include the weight of such immediate wrapper if the total weight of such immediate wrapper does not exceed :

(a) Eight per cent, Where such immediate wrapper is a waxed paper or other paper with wax or aluminium foil under strip; or

(b) Six per cent, In case of other paper of the total net weight of all the items of confectionery contained in the package minus the weight of immediate wrapper.

8. Lot/Code/Batch identification

A batch number or code number or lot number which is a mark of identification by which the food can be traced in the manufacture and identified in the distribution, shall be given on the label.

In case of packages containing bread and milk including sterilised milk, particulars under this clause shall not be required to be given on the label.

9. Date of manufacture or packing

The date, month and year in which the commodity is manufactured, packed or pre-packed, shall be given on the label.

The month and the year of manufacture, packing or pre-packing shall be given if the "Best Before Date" of the products is more than three months. In case any package contains commodity which has a short shelf life of less than three months, the date, month and year in which the commodity is manufactured or prepared or pre-packed shall be mentioned on the label.

10. Best Before and Use By Date

(i) The month and year in capital letters upto which the product is best for consumption, in the following manner, namely:

"BEST BEFORE. MONTHS AND YEAR

OR

"BEST BEFORE. MONTHS FROM PACKAGING

OR

"BEST BEFORE.MONTHS FROM MANUFACTURE

(Note: — blank be filled up)

(ii) In case of package or bottle containing sterilised or Ultra High Temperature treated milk, soya milk, flavoured milk, any package containing bread, dhokla, bhelpuri, pizza, doughnuts, khoa, paneer, or any uncanned package of fruits, vegetable, meat, fish or any other like commodity, the declaration be made as follows:

"BEST BEFORE ....DATE/MONTH/YEAR"

OR

"BEST BEFORE.DAYS FROM PACKAGING"

OR

"BEST BEFORE ....DAYS FROM MANUFACTURE"

Note:

(a) Blanks to be filled up

(b) Month and year may be used in numerals

(c) Year may be given in two digits

(iii) On packages of Aspartame, instead of Best Before date, Use by date/ recommended last consumption date/expiry date shall be given, which shall not be more than three years from the date of packing;

(iv) In case of infant milk substitute and infant foods instead of Best Before date, Use by date/recommended last consumption date/expiry date shall be given,

(v) Provided further that the declaration of best before date for consumption shall not be applicable to:

(i) wines and liquors

(ii) alcoholic beverages containing 10 percent or more by volume of alcohol.

Provided further that above provisions except net weight/net content, nutritional information, manufacturer's name and address, date of manufacture and "best before" shall not apply in respect of carbonated water (plain soda and potable water impregnated with carbon dioxide under pressure) packed in returnable glass bottles.

11. Country of origin for imported food:

(i) The country of origin of the food shall be declared on the label of food imported into India.

(ii) When a food undergoes processing in a second country which changes its nature, the country in which the processing is performed shall be considered to be the country of origin for the purposes of labelling.

12. Instructions for use

(i) Instructions for use, including reconstitution, where applicable, shall be included on the label, if necessary, to ensure correct utilization of the food, part of label depicting instruction for use is displayed in Figure 11.7.

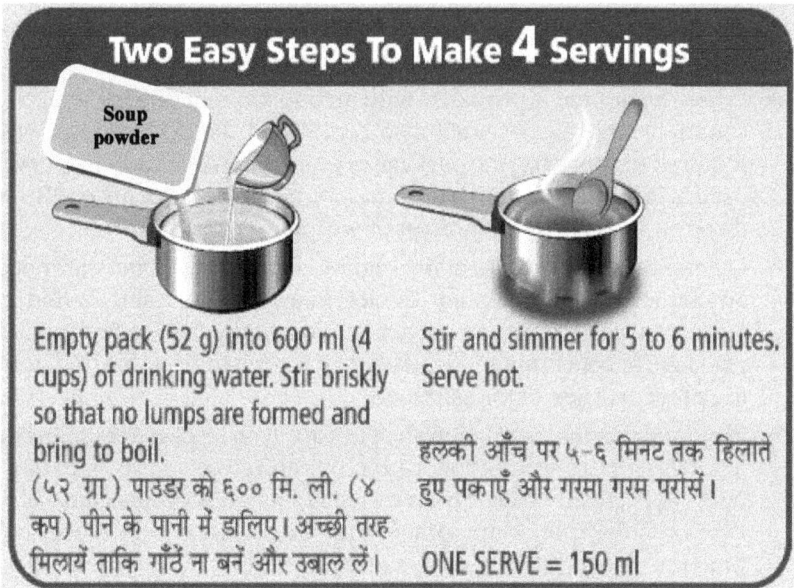

Two Easy Steps To Make 4 Servings

Soup powder

Empty pack (52 g) into 600 ml (4 cups) of drinking water. Stir briskly so that no lumps are formed and bring to boil.
(५२ ग्रा) पाउडर को ६०० मि. ली. (४ कप) पीने के पानी में डालिए। अच्छी तरह मिलायें ताकि गाँठें ना बनें और उबाल लें।

Stir and simmer for 5 to 6 minutes. Serve hot.

हल्की आँच पर ५-६ मिनट तक हिलाते हुए पकाएँ और गरमा गरम परोसें।

ONE SERVE = 150 ml

Figure 11.7: Part of a Label Showing the Instruction.

11.3.4 Manner of Declaration

11.3.4.1 General Conditions

1. Any information or pictorial device written, printed, or graphic matter may be displayed in the label provided that it is not in conflict with the requirements of these regulations.

2. Every declaration which is required to be made on package under these regulations shall be:

    (i) Legible and prominent, definite, plain and unambiguous

    (ii) Conspicuous as to size number and colour,

    (iii) As far as practicable, in such style or type of lettering as to be boldly, clearly and conspicuously present in distinct contrast to the other type, lettering or graphic material used on the package, and shall be printed or inscribed on the package in a colour that contrasts conspicuously with the background of the label.

    Further it is provided that :

    (a) Where any label information is blown, formed or moulded on a glass or plastic surface or where such information is embossed or perforated on a package, that information shall not be required to be presented in contrasting colours.

    (b) Where any declaration on a package is printed either in the form of a handwriting or hand script, such declaration shall be clear, unambiguous and legible.

3. No declaration shall be made so as to require it to be read through any liquid commodity contained in the package.

4. Where a package is provided with an outside container or wrapper, such container or wrapper shall also contain all the declarations which are required to appear on the package except where such container or wrapper itself is transparent and the declarations on the package are easily readable through such outside container or wrapper.

5. Labels not to contain false or misleading statements: A label shall not contain any statement, claim, design, device, fancy name or abbreviation which is false or misleading in any particular concerning the food contained in the package, or concerning the quantity or the nutritive value or in relation to the place of origin of the said food.

    Provided that this regulation shall not apply in respect of established trade or fancy names of confectionery, biscuits and sweets, such as, barley, sugar, bull's eye, cream cracker or in respect of aerated waters, such as, Ginger Beer or Gold-Spot or any other name in existence in international trade practice.

11.3.4.2 Principal Display Panel

The information required under these Regulations shall be given on the principal display panel of the package or container and such information may be given in the following manner.

    (a) All information should be grouped together and given at one place.

        OR

    The pre-printed information be grouped together and given in one place and,

(b) Online information or those not pre-printed be grouped together in another place.

1. Area of the principal display panel

The area of principal display panel shall not be less than —

(a) In the case of a rectangular container, forty percent of the product of height and width of the panel of such container having the largest area;

(b) In case of cylindrical or nearly cylindrical, round or nearly round, oval or nearly oval container, twenty percent of the product of the height and average circumference of such container; or

(c) In the case of container of any other shape, twenty percent of the total surface area of the container except where there is label, securely affixed to the container, such label shall give a surface area of not less than ten percent of the total surface area of the container.

In the case of package having a capacity of five cubic centimeters or less, the principal display panel may be card or tape affixed firmly to the package or container and bearing the required information under these regulations.

11.3.4.3 The Height of Numercial in the Declaration

(i) The height of any numeral required under these regulations, on the principal display panel shall not be less than-

(a) As shown in Table 11.5, if the net quantity is declared in terms of weight or volume, and

(b) As shown in Table 11.6, if the net quantity is declared in terms of length, area or number.

Table 11.5: Height of Numeral when Net Quantity is in Weight or Volume

Sl.No.	Weight/Volume	Minimum Height of Numeral in mm	
		Normal Case	When Blown, Formed Moulded, or Perforated on Container
1.	Upto 50g/ml	1	2
2.	Above 50g/ml upto 200g/ml	2	4
3.	Above 200 g/ml upto 1 kg/litre	4	6
4.	Above 1 kg/litre	6	8

(ii) The height of letters in the declaration under section 11.3 shall not be less than 1 mm height when blown, formed, moulded, embossed or perforated, the height of letters shall not be less than 2mm.

Provided that the width of the letter or numeral shall not be less than one-third of its height, but this provision shall not apply in the case of numeral "I" and letters i, I and I. Provided further that in case of label declarations

Table 11.6: Height of Numeral when the Net Quantity is Declared in Terms of Length, Area or Number

Sl.No.	Weight/Volume	Minimum Height of Numeral in mm	
		Normal Case	When Blown, Formed Moulded, or Perforated on Container
1.	Upto 100cms square	1	2
2.	Above 100cms square upto 500cms square	2	4
3.	Above 500cms square upto 2500cms square	4	6
4.	Above 2500cms square	6	8

required under section 11.3.5 except in case declaration specifying instructions for use or preparation of the product, the size of letters shall not be less than 3mm.

11.3.5 Specific Requirements/Restrictions on Manner of Labelling

11.3.5.1: Labelling of Infant Milk Substitute and Infant Food

1. An article of infant milk substitutes/infant foods, whose standards are not prescribed under Food Safety and Standards (Food Products standards and Food Additives) Regulations, 2011 shall be manufactured for sale, exhibited for sale or stored for sale only after obtaining the approval of such articles of food and its label from the FSSAI.

2. Without prejudice to any other provisions relating to labelling requirements contained in these regulations, every container of infant milk substitute or infant food or any label affixed thereto shall indicate in a clear, conspicuous and in an easily readable manner, the words "IMPORTANT NOTICE" in capital letters and indicating there under the following particulars, namely:

   (i) a statement "MOTHER'S MILK IS BEST FOR YOUR BABY" in capital letters. The types of letters used shall not be less than five millimeters and the text of such statement shall be in the Central Panel of every container of infant milk substitute or infant food or any label affixed thereto. The colour of the text printed or used shall be different from that of the background of the label, container as the case may be. In case of infant food, a statement indicating "infant food shall be introduced only (after the age of six months and upto the age of two years)" shall also be given;

   (ii) a statement that infant milk substitute or infant food should be used only on the advice of a health worker as to the need for its use and the proper method of its use;

(iii) a warning that infant milk substitute or infant food is not the sole source of nourishment of an infant;

(iv) a statement indicating the process of manufacture (e.g spray dried) except in case of infant foods, instruction for appropriate and hygienic preparation including cleaning of utensils, bottles and teats and warning against health hazards of inappropriate preparations, as under; "Warning/caution-Careful and hygienic preparation of infant foods/infant milk substitute is most essential for health. Do not use fewer scoops than directed since diluted feeding will not provide adequate nutrients needed by your infant. Do not use more scoops than directed since concentrated feed will not provide the water needed by your infant".

(v) the approximate composition of nutrients per 100 gms. of the product including its energy value in Kilo Calories/Joules;

(vi) the storage condition specifically stating "store in a cool and dry place in an air tight container" or the like (after opening use the contents within the period mentioned or the expiry date whichever is earlier);

(vii) the feeding chart and directions for use and instruction for discarding leftover feed;

(viii) Instruction for use of measuring scoop (level or heaped) and the quantity per scoop (scoop to be given with pack);

(ix) indicating the Batch No. Month and Year of its manufacture and expiry date;

(x) the protein efficiency ratio (PER) which shall be minimum 2.5 if the product other than infant milk substitute is claimed to have higher quality protein;

(xi) the specific name of the food additives, if permitted, shall be declared in addition to appropriate class names.

3. No containers or label referred to in **11.3.5.1(2)** relating to infant milk substitute or infant food shall have a picture of infant or women or both. It shall not have picture or other graphic materials or phrases designed to increase the saleability of the infant milk substitute or infant food. The terms "Humanised" or "Maternalised" or any other similar words shall not be used. The Package and/or any other label of infant milk substitute or infant food shall not exhibit the words, "Full Protein Food", "energy Food", "Complete food" or "Health Food" or any other similar expression.

4. The containers of infant milk substitute meant for (premature baby (born before 37 weeks)/low birth weight infant (less than 2500gm) or labels affixed thereto shall indicate the following additional information, namely:

(i) the words [PREMATURE BABY (BORN BEFORE 37 WEEKS) LOW BIRTH WEIGHT (LESS THAN 2.5 KG)] in capital letters along with the product name in central panel;

(ii) a statement "the low birth weight infant milk substitute shall be withdrawn under medical advice as soon as the mother's milk is sufficiently available'; and

(iii) a statement "TO BE TAKEN UNDER MEDICAL ADVICE" in capital letters.

5. The product which contains neither milk nor any milk derivatives shall be labelled "contains no milk or milk product" in conspicuous manner.

6. The container of infant milk substitute for lactose or lactose and sucrose intolerant infants or label affixed thereto shall indicate conspicuously "LACTOSE-FREE or SUCROSE-FREE or LACTOSE and SUCROSE-FREE" in capital letters and statement "TO BE TAKEN UNDER MEDICAL ADVICE" and shall also bear the following statements, namely:

"Lactose free Infant Milk Substitute should only be used in case of diarrhea due to lactose intolerance.

The lactose free/sucrose free Infant Milk Substitute should be withdrawn if there is no improvement in symptoms of intolerance".

7. The container of infant milk substitute meant for infants with allergy to cow's/buffalo's milk protein or soy protein or label affixed thereto shall indicate conspicuously "HYPOALLERGENIC FORMULA" in capital letters and statement "TO BE TAKEN UNDER MEDICAL ADVICE".

8. Declaration to be surrounded by line:

There shall be a surrounding line enclosing the declaration where the words "unsuitable for babies" are required to be used.

Distance of surrounding line: The distance between any part of the words "unsuitable for babies" surrounding the line enclosing these words shall not be less than 1.5 mm.

11.3.5.2 Labelling of Edible Oils and Fats

1. The package, label or the advertisement of edible oils and fats shall not use the expressions "Super- Refined", "Extra-Refined", "Micro-Refined", "Double-Refined", Ultra-Refined", "Anti-Cholesterol", "Cholesterol Fighter", "Soothing to Heart", "Cholesterol Friendly", "Saturated Fat Free" or such other expressions which are an exaggeration of the quality of the Product.

2. Every container in which solvent-extracted oil or de-oiled meal or edible flour is packed for sale shall, at the time of sale by the producer, bear the following particulars in English or Hindi (Devnagri script):

(i) the name, trade name, if any, or description of the solvent-extracted oil or de-oiled meal or edible flour, as the case may be:

(ii) in the case of oil not conforming to the standards of quality for "refined" grade solvent extracted oils specified in regulation 2.2.6 (1) of Food Safety and Standards (Food Products Standards and Food Additive)

Regulation, 2011 for Edible vegetable oil/Vanaspati, a declaration in a type-size of not less than 50 mm, as follows shall appear on the label:

(a) "NOT FOR DIRECT EDIBLE CONSUMPTION", in the case of oils complying with the requirements for the "semi-refined" or "raw-grade 1" grades of oil specified in regulation 2.2.6 (1) of Food Safety and Standards(Food Products standards and Food Additive) Regulation, 2011.

(b) "FOR INDUSTRIAL NON-EDIBLE USES ONLY", in the case of oils not complying with the requirements under item (a) above;

(iii) the name and business particulars of the producer;

(iv) the net weight of the contents in the container;

(v) the batch number, month and year of manufacture.

When solvent extracted oils are transported in bulk in rail tank-wagons or road tankers, or where de-oiled meal or edible flour is transported in bulk either for storage in silos or transferred to ship for bulk shipment, it shall be sufficient if the aforesaid particulars are furnished in the accompanying documents.

3. Every container in which solvent is packed shall, at the time of sale by the manufacturer or dealer thereof, bear the Indian Standards Institution certification mark.

4. Every container in which vanaspati, margarine, bakery shortening, blended edible vegetable oils, mixed fat spread and refined vegetable oil is packed in addition to other labelling requirements provided in these regulations shall bear the following particulars in English or Hindi in Devnagri script:

(a) The name/description of the contents, "free from Argemone Oil";

(b) The mass/volume of the contents;

5. Every container of refined vegetable oil shall bear the following label, namely:

| Refined (name of the Oil) Oil |

The container of imported edible oil shall also bear the word, "Imported", as prefix.

6. Every package containing an admixture of palmolein with groundnut oil shall carry the following label, namely,

| BLEND OF PALMOLEIN AND GROUNDNUT OIL |
| Palmolein........ per cent |
| Groundnut oil........ per cent |

7. Every package containing an admixture of imported rape-seed oil with mustard oil, shall carry the following label, namely :

> BLEND OF IMPORTED RAPE-SEED OIL AND MUSTARD OIL
>
> Imported rape-seed oil........per cent
>
> Mustard oil........per cent

8. Every package of vanaspati made from more than 30 percent of Rice bran oil shall bear the following label, namely :

> This package of vanaspati is made from more than
> 30 per cent Rice bran oil by weight

9. Every package containing Fat Spread shall carry the following labels namely:

9.1 Milk Fat Spread

Use before	. . . .
Date of packing	. . . .
Total Milk Fat Content Per cent by weight	. . . .

9.2 Mixed Fat Spread

Use before	. . . .
Date of packing	. . . .
Per cent by weight	. . . .
Milk Fat Content	. . . .
Total Milk Fat Content Per cent by weight	. . . .
Vegetable Fat Spread Use before	. . . .
Date of packing	. . . .
Total Fat Content Per cent by weight	. . . .

10. A package containing annatto colour in vegetable oils shall bear the following label namely :

> Annatto colour in oil (Name of oil/oils) used

11. Every package containing an admixture of edible oils shall carry the following label, namely:

This blended edible vegetable oil contains an admixture of:

> (i) .... per cent by Weight
> (ii) .... per cent by Weight

(Name and nature of edible vegetable oils *i.e.* in raw or refined form)

Date of Packing ....

There shall also be the following declaration in bold capital letters along with the name of product on front/central panel:

> NOT TO BE SOLD LOOSE

11.3.5.3 Labelling of Permitted Food Colours

1. No person shall sell a permitted synthetic food colours for use in or upon food unless its container carries a label stating the following particulars:

   (i) the words "Food Colours";
   (ii) the chemical and the common or commercial name and colour index of the dye-stuff.

2. No person or manufacturer shall sell a mixture of permitted synthetic food colours for use in or upon food unless its container carries a label stating the following particulars:

   (i) The words "Food Colour Mixture";
   (ii) The chemical and the common or commercial name and colour index of the dye stuff contained in the mixture.

3. No person shall sell a preparation of permitted synthetic food colours for use in or upon food unless its container carries a label stating the following particulars:

   (i) The words "Food Colour Preparation";
   (ii) The name of the various ingredients used in the preparation.

11.3.5.4 Labelling of Irradiated Food

Irradiated Foods–The label of a food, which has been treated with ionizing radiation, shall carry a written statement indicating the treatment in close proximity to the name of the food. In addition all packages of irradiated food shall bear the following declaration and logo, namely:

PROCESSED BY IRRADIATION METHOD

DATE OF IRRADIATION........

LICENSE NO of Irradiation Unit...............

PURPOSE OF IRRADIATION...............

11.3.5.5 Specific Labelling Requirements of Other Products

1. Coffee-Chicory Mixture

(i) Every package containing a mixture of coffee and chicory shall have affixed to it a label upon which shall be printed the following declaration:

Coffee blended with Chicory

This mixture contains

Coffee................................ Per cent

Chicory.............................. Per cent

(ii) Every package containing Instant Coffee-Chicory mixture shall have affixed to it a label upon which shall be printed the following declarations;

Instant Coffee-Chicory mixture made from blends of coffee and chicory

Coffee................................ Per cent

Chicory.............................. Per cent

2. Condensed Milk or Desiccated (Dried) Milk

Every package containing condensed milk or desiccated (dried) milk shall bear a label upon which is printed such one of the following declarations as may be applicable or such other declaration substantially to the like effect as may be allowed by the State Government, namely,

(i) In the case of condensed milk (unsweetened)

CONDENSED MILK UNSWEETENED

(Evaporated Milk) (This tin contains the equivalent) of (x)....litres of toned milk

(ii) In the case of condensed milk (sweetened):

CONDENSED MILK SWEETENED

This tin contains the equivalent of (x)...... litres of toned milk with sugar added

(iii) In the case of condensed skimmed milk (unsweetened):

CONDENSED SKIMMED MILK UNSWEETENED

(Evaporated Skimmed Milk) This tin contains the equivalent of (x)........ litres of skimmed milk

(iv) In the case of condensed skimmed milk (sweetened):

CONDENSED SKIMMED MILK SWEETENED

This tin contains the equivalent of (x)...litres of skimmed milk with sugar added"

(v) In the case of condensed milk (sweetened and flavoured):

This has been flavoured with......................

NOT TO BE USED FOR INFANTS BELOW SIX MONTHS

(vi) In the case of condensed milk/condensed Skimmed milk (unsweetened) Sterilised by Ultra High Temperature (UHT) treatment:

This has been sterilised by UHT Process

(vii) In the case of milk powder:

MILK POWDER

This tin contains the equivalent of (x)........ litres of toned milk

(viii) In the case of milk powder which contains lecithin:

MILK POWDER IN THIS PACKAGE CONTAINS LECITHIN

(ix) In the case of partly skimmed milk powder :

PARTLY SKIMMED MILK POWDER

This tin contains the equivalent of (x). . . . . litres of
partly skimmed milk having.per cent milk fat

(x) In the case of skimmed milk powder:

SKIMMED MILK POWDER

This tin contains the equivalent of (x). . . . . litres of skimmed milk

3. The declaration shall in each case be completed by inserting at (x) the appropriate number in words and in figures, for example, "one and a half (1½)", any fraction being expressed as eight quarters or a half, as the case may be.

4. There shall not be placed on any package containing condensed milk or desiccated (dried) milk any comment on, explanation of, or reference to either the statement of equivalence, contained in the prescribed declaration or on the words "machine skimmed" "skimmed" or "unsuitable for babies" except instructions as to dilution as follows:

    "To make a fluid not below the composition of toned milk or skimmed milk (as the case may be) with the contents of this package, add (here insert the number of parts) of water by volume to one part by volume of this condensed milk or desiccated (dried) milk".

    Sweetened condensed milk and other similar products which are not suitable for infant feeding shall not contain any instruction of modifying them for infant feeding.

5. Wherever the word "milk" appears on the label of a package of condensed skimmed milk or of desiccated (dried) skimmed milk as the description or part of the description of the contents, it shall be immediately preceded or followed by the word "machine skimmed" or "partly skimmed", as the case may be.

6. Fluid Milk

The caps of the milk bottles/pouch/tetrapack shall clearly indicate the nature of the milk contained in them. The indication may be either in full or by abbreviation shown below :

(i) Buffalo milk may be denoted by the letter 'B'.

(ii) Cow milk may be denoted by the letter 'C'

(iii) Goat milk may be denoted by the letter 'G'

(iv) Standardized milk may be denoted by the letter 'S'

    (v) Toned milk may be denoted by the letter 'T'

    (vi) Double toned milk may be denoted by the letter 'DT'

    (vii) Skimmed milk may be denoted by the letter 'K'

    (viii) Pasteurised milk may be denoted by the letter 'P; followed by the class of milk. For example Pasteurised Buffalo milk shall bear the letters 'PB '.

Alternatively suitable indicative colours of the packs/caps/bags shall be indicative of the nature of milk contained in them, the classification of colours being displayed at places where milk is sold\stored or exhibited for sale, provided that the same had been simultaneously intimated to the concerned Designated Officer, and information disseminated through the local media.

7. Ice-cream

Every dealer in ice-cream or mixed ice-cream who in the street or other place of public resort, sells or offers or exposes for sale, ice-cream or ice-candy, from a stall or from a cart, barrow or other vehicle or from a basket, phial, tray or other container used without a staff or a vehicle shall have his name and address along with the name and address of the manufacturer, if any, legibly and conspicuously 'displayed' on the stall, vehicle or container as the case may be.

8. Hingra

Every container containing Hingra shall bear a label upon which is printed a declaration in the following form, namely,

"This container contains Hingra (Imported from Iran\Afghanistan) and is certified to be conforming to the standards laid down in the Food Safety and Standards regulations"

9. Light Black Pepper

Every package containing light black pepper shall bear the following label in addition to the Agmark seal and the requirements prescribed under regulation 11.3.2 and 11.3.3 of these regulations:

> Light Black Pepper (Light berries)

10. Cassia Bark

Every package containing "Cassia Bark" shall bear the following label:

> CASSIA BARK (TAJ)

11. Cinnamon

Every package containing "CINNAMON" shall bear the following label:

> CINNAMON (DALCHINI)

12. Chillies Containing Added Edible Oil

Every package of chillies which contains added edible oil shall bear the following label:

> CHILLIES IN THIS PACKAGE CONTAINS AN ADMIXTURE OF NOT MORE THAN 2 PERCENT OF........(NAME OF OIL) EDIBLE OIL

13. Ice-cream, Kulfi, Kulfa and Chocolate Ice-cream Containing Starch

Every package of ice-cream, kulfi, kulfa and chocolate ice-cream containing starch shall have a declaration on a label as specified in regulation **11.3.5.9** (2)

14. Masala

Every package of mixed masala fried in oil shall bear the following label:

> MIXED MASALA (FRIED) THIS MASALA HAS BEEN FRIED IN
> (Name of the edible oil used)

15. Compounded Asafoetida

Every container of compounded asafoetida shall indicate the approximate composition of edible starch or edible cereal flour used in the compound, on the label.

16. Maida Treated with Improver or Bleaching Agents

Every package containing maida treated with improver or bleaching agents shall carry the following label, namely,-

> WHEAT FLOUR TREATED WITH IMPROVER/BLEACHING AGENTS, TO BE USED BY BAKERIES ONLY

17. Malted Milk Food Containing Added Natural Colouring Matter

Unless otherwise provided in these regulations, every package of malted milk food which contains added natural colouring matter except caramel, shall bear the following label, namely,-

> MALTED MILK FOOD IN THIS PACKAGE CONTAINS PERMITTED NATURAL COLOURING MATTER

18. Food Containing Added Monosodium Glutamate

Every advertisement for and/or a package of food containing added Monosodium Glutamate shall carry the following declaration, namely,-

> This package of (name of the food contains added)......... MONOSODIUM GLUTAMATE NOT RECOMMENDED FOR INFANTS BELOW -12 MONTHS

19. Refined Salseed Fat

Every container of refined salseed fat shall bear the following label, namely:

> REFINED SALSEED FAT FOR USE IN BAKERY AND CONFECTIONERY ONLY

20. Salt

Every container or package of table iodised salt or iron fortified common salt containing permitted anticaking agent shall bear the following label, namely,-

> IODIZED SALT/IRON FORTIFIED COMMON SALT* CONTAINS PERMITTED ANTICAKING AGENT

* Strike out whichever is not applicable

21. Every container or package of iron fortified common salt shall bear the following label, namely,

> IRON FORTIFIED COMMON SALT

22. Dried Glucose Syrup Containing Sulphur Dioxide

Every package of Dried Glucose Syrup containing sulphur dioxide exceeding 40 ppm shall bear the following label namely,

> DRIED GLUCOSE SYRUP FOR USE IN SUGAR CONFECTIONERY ONLY

23. Tea with Added Flavour

A package containing tea with added flavour shall bear the following label, namely,

> "FLAVOURED TEA" (common name of permitted flavour/percentage) Registration No........

24. Food Containing Artifical Sweetner

Every package of food which is permitted to contain artificial sweetener mentioned in table given in regulation 3.1.3 (1) of Food Safety and standards (Food

Products standards and Food Additive) Regulations, 2011 and an advertisement for such food shall carry the following label, namely,

---

(i) This contains........(Name of the artificial sweeteners).

(ii) Not recommended for children.

(iii) (a) *Quantity of sugar added........gm/100 gm.

    (b) No sugar added in the product.

(iv) *Not for Phenylketonurics (if Aspartame is added)

---

*strike out whatever is not applicable

25. In addition to the declarations under regulation **11.3.5.5** (24 and 26), every package of food which is permitted to contain **artificial sweetener** mentioned in table in regulation 3.1.3 (1)of Food Safety and Standards (Food Products standards and Food Additive) Regulations, 2011 and an advertisement for such food shall carry the following label, namely,

---

CONTAINS ARTIFICIAL SWEETENER AND FOR CALORIE CONSCIOUS

The declaration under regulation **11.3.5.5** (25) shall be provided along with name or trade name of product and shall be half of the size of the name/trade name. The declaration may be given in two sentences, but in the same box.

26. Artificial Sweetner

Every package of Aspertame (Methyl ester), Acesulfame K, Sucralose and Saccharin Sodium, Neotame marketed as Table Top Sweetener and every advertisement for such Table Top Sweetener shall carry the following label, namely,-

---

(i) Contains........ (name of artificial sweetener)

(ii) Not recommended for children

The package of aspartame (Methyl ester), marketed as Table Top Sweetener and every advertisement for such Table Top Sweetener shall carry the following label, namely,

---

"Not for Phenylketonurics"

---

27. Every package of food which is permitted to contain a mixture of Aspartame (Methyl Ester) and Acesulfame Potassium Sweeteners mentioned in the Table given in regulation 3.1.3(1) of Food Safety and Standards (Food

Products Standards and Food Additive) Regulation,2011, shall carry the following label, namely,

---

This........(Name of food) contains .... contains an admixture of Aspartame (Methyl Ester and Acesulfame Potassium.

Not recommended for children.

(a) *Quantity of sugar added........gm/100gm,

(b) No sugar added in the product........

*Not for Phenylketoneurics (if Aspartame is added)

---

*strike out whatever is not applicable

28. Food Containing a Mixture of Acesulfame Potassium and Sucralose Sweeteners

Every package of food which is permitted to contain a mixture of Acesulfame Potassium and Sucralose sweeteners mentioned in the Table given in Regulation 3.1.2 (1) of Food Safety and Standards (Food Products Standards and Food Additive) Regulation,2011 shall carry the following label, namely,

---

(i) This........(Name of Food) contains a mixture of Sucralose and Acesulfame Potassium;

(ii) Not recommended for children;

(iii) *(a) Quantity of sugar added........gm/100gm;*

    (b) No sugar added in the product.

(*Strike out whichever is not applicable)

29. Pan Masala

Every package of Pan Masala and advertisement relating thereto, shall carry the following warning, namely,

---

Chewing of Pan Masala is injurious to health

---

30. Supari

Every package of supari and advertisement relating thereto shall carry the following warning in conspicuous and bold print, namely,-

---

Chewing of Supari is injurious to Health

---

31. *Fruit Squash*

Every package of fruit squash by whatever name it is sold, containing additional sodium or potassium salt shall bear the following label, namely,

IT CONTAINS ADDITIONAL SODIUM/POTASSIUM SALT

32. *Cheese (Hard)*

Every package of Cheese (hard), surface treated with Natamycin, shall bear the following label, namely,

SURFACE TREATED WITH NATAMYCIN

33. *Bakery and Industrial Margarine made from Rice Bran Oil*

Every package of Bakery and Industrial Margarine made from more than 30 per cent of Rice Bran Oil shall bear the following label, namely,-

This package of Bakery and Industrial Margarine is made from more than 30 per cent of Rice Bran Oil by Wt.

34. *Flavour Emulsion and Flavour Paste for Use in Beverage*

Every container or package of flavour emulsion and flavour paste meant for use in carbonated or non- carbonated beverages shall carry the following declaration, in addition to the instructions for dilution, namely,-

FLAVOUR EMULSION AND FLAVOUR PASTE FOR USE IN CARBONATED OR NON-CARBONATED BEVERAGES ONLY

35. *Drinking Water*

Every package of drinking water shall carry the following declaration in capital letters having the size of each letter as prescribed in Regulation **11.3.4.3**:

PACKAGED DRINKING WATER

One time usable plastic bottles of packaged drinking water shall carry the following declaration.

CRUSH THE BOTTLE AFTER USE

36. Mineral Water

Every package of mineral water shall carry the following declaration in capital letters having the size of each letter as prescribed in regulation **11.3.4.3**;

> NATURAL MINERAL WATER

One time usable plastic bottles of mineral water shall carry the following declaration.

> CRUSH THE BOTTLE AFTER USE

37. Food Containing Added Caffeine

Every package of food having added caffeine, shall carry the following label, namely,

> "CONTAINS CAFFEINE"

When caffeine is added in the products, it shall be declared on the body of the Container/bottle. Also in case of returnable glass bottles, which are recycled for refilling the declaration of caffeine, may be given on the crown.

38. Low Fat Paneer/Chhana

Every package of Low Fat Paneer/Chhana shall carry the following label, namely,-

> LOW FAT PANEER/CHHANA

39. Cheese

Every package of Cheese(s), if coated/packed in food grade waxes polyfilm/ wrapping of cloth, shall bear the following label, namely,

> REMOVE THE OUTER PACKING BEFORE CONSUMPTION

40. Frozen Desert/Frozen Confection

Every package of Frozen Desert/Frozen Confection shall bear the following label, namely,

> Frozen Desserts/Frozen Confection Contain........
> Milk Fat*/Edible Vegetable Oil*/and Vegetable Fat*

*strike out whatever is not applicable

41. Common Salt

Every container or package of common salt shall bear the following label, namely,

> COMMON SALT FOR IODISATION*/IRON FORTIFICATION*/
> ANIMAL USE*/PRESERVATION/MEDICINE*/INDUSTRIAL USE*

*strike out whichever is not applicable.

42. Biscuits, Bread and Cakes Containing Oligofructose

Every package of biscuits, bread and cakes containing Oligofructose shall bear the following declaration, namely,

> Contains Oligofructose (dietary fiber) —— gm/100 gm

43. Fresh Fruit Coated with Wax

Every package of fresh fruit if coated with wax shall carry the following label, namely,

> Coated with wax (give name of wax)

44. Gelatin

Gelatin meant for human consumption should be labeled as:

> "Gelatin Food Grade"

45. Food Containing Polyols

Every package of food containing Polyols shall bear the following label,

> Polyols may have laxative effects

46. Food Containing Polydextrose

Every package of food containing Polydextrose shall bear the following label:

> Polydextrose may have laxative effects

11.3.5.6 Specific Restrictions on Product Labels

(1) Labels not to contain reference to Act or rules or regulations contradictory to required particulars:

The label shall not contain any reference to the Act or any of these regulations or any comment on, or reference to, or explanation of any particulars or declaration required by the Act or any of these regulations to be included in the label which directly or by implication, contradicts, qualifies or modifies such particulars or declaration.

(2) Labels not to use words implying recommendations by medical profession:

There shall not appear in the label of any package, containing food for sale the words "recommended by the medical profession" or any words which imply or suggest that the food is recommended, prescribed, or approved by medical practitioners or approved for medical purpose.

(3) Unauthorized use of words showing imitation prohibited

   i. There shall not be written in the statement or label attached to any package containing any article of food the word 'imitation' or any word, or words implying that the article is a substitute for any food, unless the use of the said word or words is specifically permitted under these regulations.

   ii. Any fruit syrup, fruit juice, fruit squash, fruit beverages, cordial, crush or any other fruit products standardised under Food Safety and Standards (Food Products standards and Food Additives) Regulations, 2011 which does not contain the prescribed amount of fruit juice or fruit pulp or fruit content shall not be described as a fruit syrup, fruit juice, fruit squash, fruit beverages, cordial, crush or any other fruit product as the case may be.

   iii. Any food product which does not contain the specified amount of fruit and is likely to deceive or mislead or give a false impression to the consumer that the product contains fruit, whether by use of words or pictorial representation, shall be clearly and conspicuously marked on the label as **'ADDED(NAME OF THE FRUIT) FLAVOUR'.**

   iv. Any food product which contains only fruit flavours, whether natural flavours and natural flavouring substances or nature identical flavouring substances, artificial flavouring substances as single or in combination thereof, shall not be described as a fruit product and the word **"ADDED" (NAME OF FRUIT) FLAVOUR** shall be used in describing such a product;

   v. Carbonated water containing no fruit juice or fruit pulp shall not have a label which may lead the consumer into believing that it is a fruit product.

   vi. Any fruit and vegetable product alleged to be fortified with vitamin C shall contain not less than 40 mg's. of ascorbic acid per 100 gm of the product.

(4) Imitations not to be marked "pure" The word "pure" or any word or words of the same significance shall not be included in the label of a package that contains an imitation of any food.

(5) Labelling prohibitions for Drinking Water (Both Packaged and Mineral Water):

   i. No claims concerning medicinal (preventative, alleviative or curative) effects shall be made in respect of the properties of the product covered by the standard claims of other beneficial effects related to the health of the consumer shall not be made.

    ii. The name of the locality, hamlet or specified place may not form part of the trade name unless it refers to a packaged water collected at the place designated by that trade name.

    iii. The use of any statement or of any pictorial device which may create confusion in the mind of the public or in any way mislead the public about the nature, origin, composition, and properties of such waters put on sale is prohibited.

11.3.5.7 Restriction on Advertisement

There shall be no advertisement of any food which is misleading or contravening the provisions of Food Safety and Standards Act, 2006 (34 of 2006) or the rules/ regulations made thereunder.

11.3.5.8 Exemptions from Labelling Requirements

1. Where the surface area of the package is not more than 100 square centimeters, the label of such package shall be exempted from the requirements of list of ingredients, Lot Number or Batch Number or Code Number, nutritional information and instructions for use, but these information shall be given on the wholesale packages or multi piece packages, as the case may be.

2. The 'date of manufacture' or 'best before date' or 'expiry date' may not be required to be mentioned on the package having surface area of less than 30 square centimeters but these information shall be given on the wholesale packages or multipiece packages, as the case may be.

3. In case of liquid products marketed in bottles, if such bottle is intended to be reused for refilling, the requirement of list of ingredients shall be exempted, but the nutritional information specified in regulation **11.3.3** (4) these regulations shall be given on the label. Provided that in case of such glass bottles manufactured after March 19, 2009, the list of ingredients and nutritional information shall be given on the bottle.

4. In case of food with shelf-life of not more than seven days, the 'date of manufacture' may not be required to be mentioned on the label of packaged food articles, but the 'use by date' shall be mentioned on the label by the manufacturer or packer.

5. In case of wholesale packages the particulars regarding list of ingredients. Date of manufacture/packing, best before, expiry date labelling of irradiated food and, vegetarian logo/non vegetarian logo, may not be specified.

11.3.5.9 Notice on Addition, Admixture or Deficiency in Food

1. Every advertisement and every price or trade list or label for an article of food which contains an addition, admixture or deficiency shall describe the food as containing such addition, admixture or deficiency and shall also specify the nature and quantity of such addition, admixture or

deficiency and no such advertisement or price or trade list or label attached to the container of the food shall contain any words which might imply that the food is pure:

Provided that for purpose of this regulation the following shall not be deemed as an admixture or an addition, namely:

(a) salt in butter or margarine;

(b) vitamins in food.

2. Every package, containing a food which is not pure by reason of any addition, admixture or deficiency shall be labelled with an adhesive label, which shall have the following declaration:

Declaration:

This (a)........ contains an admixture/addition of not more than (b)........ per cent of (c)........

(a) Here insert the name of food.

(b) Here insert the quantity of admixture which may be present.

(c) Here insert the name of the admixture or the name of ingredient which is deficient.

Where the context demands it, the words 'contains an admixture of' shall be replaced by the words 'contains an addition of' or 'is deficient in'.

3. Unless the vendor of a food containing an addition, admixture or deficiency, has reason to believe that the purchaser is able to read and understand the declaratory label, he shall give the purchaser, if asked, the information contained in the declaratory label by word of mouth at the time of sale

4. Nothing contained in regulation **11.3.5.9** shall be deemed to authorize any person to sell any article of food required under the Act or these regulations which is to be sold in pure condition, otherwise than in its pure condition.

5. Nothing contained in regulation **11.3.5.9** shall apply in the case of sweets, confectionery, biscuits, bakery products, processed fruits, aerated water, vegetables and flavouring agents.

11.4 Global Trends (UK Traffic Light Approach)

Currently the traffic light label is used in some European countries on a voluntary basis. This can be called a new technique to compare a products nutrition with reference intake traffic signal warning. Under this approach the nutritional labeling can be done based on the colour so that all consumers can understand what they are eating. A traffic light labelling system can be used as a key to help people to make informed choices while choosing to eat prepacked food. Traffic light labels provide a clear indication about the level of fat, saturated fat, sugar and salt in products that are highly processed, like ready to eat meals. There are three colours used in this approach:

☆ **Red - high** - enjoy it ONCE in a while

☆ **Amber - medium** - OK most of the time

☆ **Green - low** - GO for it

These could be servings/energy/amount of nutrient being provided by product.

Food sold prepacked may be labelled with a **traffic light label** showing at a glance the proportions of fat, saturated fats, sugar, and salt using traffic light signals for high (*red*), medium (*amber*) and low (*green*) percentages for each of these ingredients. Foods with 'green' indicators are healthier and to be preferred over those with 'red' ones. The label is on the front of the package and easier to mark and interpret than nutritional labelling. The nutritional information is difficult to understand for many individuals including children, and is not easy to do quick comparisons. The use of traffic light labelling is supported by many physician groups including the British Medical Association and welcomed by consumers. Despite worries from some in the food industry that red foods would be shunned, the British Medical Association, Food Standards Agency and others agree that consumers interpret the labels sensibly, realise they can have red foods as a 'treat'. Following are the indications for labels:

Food (per 100g)

Substance	Green (Low)	Amber (Medium)	Red (High)
Fat	less than 3g	between 3g and 20g	more than 20g
Saturated fats	less than 1.5g	between 1.5g and 5g	more than 5g
Sugar	less than 5g	between 5g and 12.5g	more than 12.5g
Salt	less than 0.3g	between 0.3g and 1.5g	more than 1.5g

Drinks (per 100ml)

Substance	Green (Low)	Amber (Medium)	Red (High)
Fat	less than 1.5g	between 1.5g and 10g	more than 10g
Saturated fats	less than 0.75g	between 0.75g and 2.5g	more than 2.5g
Sugar	less than 2.5g	between 2.5g and 6.3g	more than 6.3g
Salt	less than 0.3g	between 0.3g and 1.5g	more than 1.5g

An example of label with front-of-pack nutrition labels using red, amber and green colour-coding is shown below:

11.5 Some Recent Developments in Labelling in India

1. Labelling Genetically Modified Foods

In India Ministry of Agriculture announced in year 2000 that genetically engineered seeds and foods would not be allowed in our country until their safety is scientifically proven. Presently it is illegal to import, store, manufacture or sell any genetically modified food in India.

2. Labelling of Nutraceuticals

Till date there is not well defined regulations guiding the labeling of neutraceutical product, presently they have to adhere to the guidelines of FSSA. It is important to make a law so that companies don't make misleading claims for neutraceutical food products. For the food products proposing health claim, a number of factors have to be taken into consideration which includes benefit to risk analysis, evaluation of efficacy, toxicity and health regulations particularly in case of functional foods.

Development and regulation of functional foods initiated in Japan in the early eighties with advances in chemical identication of bioactive compounds, processing and formulation of foods as well as analysing the molecular mechanisms involved in the modulation of metabolic disorders. The initial regulatory background for functional foods was established by Japan in 1991 with the introduction of "foods for specied health use" (**FOSHU**) policy that facilitates production and marketing of health-promoting foods. Since 1991 over 600 FOSHU products are now available in the Japanese market. The initiative taken by Japanese regulatory authority has encouraged growth in the global functional foods market especially in the USA, European Union, and Canada, all of which now have various regulatory bodies to govern the manufacture and marketing of health-promoting food products.

References

1. FSSAI
2. http://www.nhs.uk/Livewell/Goodfood/Pages/food-labelling.aspx#Tr
3. http://www.mixph.com/2008/04/the-importance-of-food-labeling.html
4. http://www.nhs.uk/Livewell/Goodfood/Documents/Eatwellplate.pdf
5. Food Science, fifth edition by B. Srilakshmi, New Age Publisher.
6. Functional Foods and Nutraceuticals by Rotimi E. Aluko, Springer, 2012
7. Food Science, 5th edition, by Norman N. Potter, Joseph H. Hotchkiss. CBS Publishers and Distributors
8. www.bhf.org.uk/get-involved/campaigning/food-labelling.aspx
9. http://en.wikipedia.org/wiki/Traffic_light_rating_system

Labels of different products available in the market were taken and part of the label have been used in the chapter for teaching purpose only.

Index

www.ingramcontent.com/pod-product-compliance
Lightning Source LLC
Chambersburg PA
CBHW021436180326
41458CB00001B/291